Forestry Applications

In 2012, a Forestry Special Interest Group (FSIG) was founded within the Canadian Operational Research Society (CORS). Besides a general commitment to promoting the application of operational research (OR) to forest management and forest products industry problems, the FSIG has two concrete mandates: organizing the forestry cluster at the annual CORS conference and managing the editorial process for forestry-themed special issues of *INFOR*. The FSIG has been very successful in the first of these two mandates, with record attendance at the forestry cluster, the hosting of several special sessions, financial and in-kind support from the NSERC Strategic Network on Value Chain Optimization (VCO), and the inauguration of the David Martell Student Paper Prize in Forestry (DMSPPF). This is the first compilation of forestry-themed papers since the inauguration of the CORS FSIG. The six pieces selected for the special issue, now published as a book, feature applications of OR to a wide range of forest management and forest products industry contexts, including supply-chain planning, lumber production planning, demand-driven harvest and transportation planning, and fire-aware wood supply planning.

This book was originally published as a special issue of the *INFOR: Information Systems and Operational Research* journal.

Gregory Paradis is a forest engineer working on linking forest ecosystem management and forest products supply-chain planning. He is currently affiliated with the Department of Forest Resources Management at the University of British Columbia, Canada.

Forestry Applications

Edited by
Gregory Paradis

Routledge
Taylor & Francis Group

LONDON AND NEW YORK

First published 2018
by Routledge
2 Park Square, Milton Park, Abingdon, Oxon, OX14 4RN, UK

and by Routledge
52 Vanderbilt Avenue, New York, NY 10017

First issued in paperback 2020

Routledge is an imprint of the Taylor & Francis Group, an informa business

British Library Cataloguing-in-Publication Data
A catalogue record for this book is available from the British Library

ISBN 13: 978-0-367-57180-1 (pbk)
ISBN 13: 978-1-138-57639-1 (hbk)

Typeset in Minion Pro
by codeMantra

Publisher's Note
The publisher accepts responsibility for any inconsistencies that may have arisen during the conversion of this book from journal articles to book chapters, namely the possible inclusion of journal terminology.

Disclaimer
Every effort has been made to contact copyright holders for their permission to reprint material in this book. The publishers would be grateful to hear from any copyright holder who is not here acknowledged and will undertake to rectify any errors or omissions in future editions of this book.

Contents

Citation Information

The chapters in this book were originally published in *INFOR: Information Systems and Operational Research*, volume 54, issue 3 (August 2016). When citing this material, please use the original page numbering for each article, as follows:

Chapter 5

Development of a threat index to manage timber production on flammable forest land-scapes subject to spatial harvest constraints
Juan J. Troncoso, Andrés Weintraub and David L. Martell
INFOR: Information Systems and Operational Research, volume 54, issue 3 (August 2016)
pp. 262–281

Chapter 6

A model approach to include wood properties in log sorting and transportation planning
Gert Andersson, Patrik Flisberg, Maria Nordström, Mikael Rönnqvist and Lars Wilhelmsson
INFOR: Information Systems and Operational Research, volume 54, issue 3 (August 2016)
pp. 282–303

For any permission-related enquiries please visit:
http://www.tandfonline.com/page/help/permissions

INTRODUCTION

Editorial note for the special issue on forestry applications

In 2012, a Forestry Special Interest Group (FSIG) was founded within the Canadian Operational Research Society (CORS). Besides a general commitment to promoting the application of operational research (OR) to forest management and forest products industry problems, the FSIG has two concrete mandates: organizing the forestry cluster at the annual CORS conference, and managing the editorial process for forestry-themed special issues of *INFOR*.

The FSIG has been very successful in the first of these two mandates, with record attendance at the forestry cluster over the last four years, hosting of several special sessions, financial and in-kind support from the NSERC Strategic Network on Value Chain Optimisation (VCO), and the inauguration of the David Martell Student Paper Prize in Forestry (DMSPPF).

This special issue is the first compilation of forestry-themed papers since the inauguration of the CORS FSIG. The six papers selected for this special issue feature applications of OR to a wide range of forest management and forest products industry contexts, including supply-chain planning, lumber production planning, demand-driven harvest and transportation planning, and fire-aware wood supply planning.

Alayet, Lehoux, Lebel and Bouchard model weekly procurement and production activities for a network of forest products companies, over a one-year planning horizon, assuming centralized network planning. Total network profit is maximized by solving a mixed-integer linear program. Their introduction of explicit wood-fibre freshness constraints in this context is novel. Using a hypothetical supply chain case, they analyze a variety of scenarios under different demand, price and fibre aging assumptions. They show that price, demand, and fibre freshness all impact total profit of the network of forest products companies.

Lehoux, Lebel and Elleuch study potential benefits of collaborative planning for a network of five sawmills supplying chips to a single paper mill. They model fibre flows through the network by solving a linear network flow model, assuming centralized network planning. They simulate the impact of three supply coordination strategies on per-business-unit and total network profit, and show significant potential improvement in total network profit under collaborative planning. Finally, they apply three game-theoretic benefit-sharing approaches to explore relative stability of the collaborative scenarios.

Azevedo, D'Amours and Rönnqvist propose advances to the allocated Available-to-Promise (aATP) order promising mechanism, in a softwood lumber manufacturing context. They extend the state-of-the-art aATP by integrating a network perspective, multiple products per order, dynamic market price, and test the usefulness of aATP under high levels of demand and price uncertainty. They test their enhanced order-promising mechanism using real data from a Canadian softwood lumber manufacturer, and show that their proposed approach can improve profitability.

Rafiei, Nourelfath, Gaudrault, Santa-Eulalia and Bouchard examine the impact of uncontrollable supply on the profitability of a Canadian wood remanufacturing mill. They use a combined simulation/optimization framework to analyze the impact of perturbations in target supply quality on the performance of the remanufacturing business. They compile these results in terms of key performance indicator (KPI) thresholds, which can help managers make more informed decisions regarding supply management policy.

Andersson, Flisberg, Nordström, Rönnqvist and Wilhelmsson integrate a wood quality term into a combined log sorting and transportation planning optimization model. The efficiency and value-creation potential of sawmills is closely linked to the quality of logs used as raw material. By explicitly considering wood quality in their tactical-scale log procurement model, the authors propose to evaluate the tradeoff relationship between procurement cost and added value at the sawmill. They test their model on a synthetic Swedish dataset, and present results for a number of scenarios.

Troncoso, Weintraub and Martell incorporate a stand-level fire threat index into a harvest scheduling model. They use a bonus term in the objective function to steer harvesting towards high-fire-risk stands. Using a synthetic dataset, they show that the inclusion of a fire-threat index in the wood supply model can increase total volume harvested, increase terminal volume at the end of the planning horizon, and decrease area burned.

I would like to thank all the authors who submitted their work, as well as the reviewers who contributed their valuable time to help make this special issue possible. Finally, I would like to thank the past *INFOR* Editor-in-Chief, Bernard Gendron, and the current Co-Editors-in-Chief, Elkafi Hassini and Samir Elhedhli, for their support, advice, and patience in arranging for this special issue of *INFOR*.

Gregory Paradis
Guest Editor
iD http://orcid.org/0000-0001-9618-8797

Centralized supply chain planning model for multiple forest companies

C. Alayet, N. Lehoux, L. Lebel and M. Bouchard

ABSTRACT

In this paper, we present a mathematical model to plan logistics activities in a forest products supply chain. In particular, a mixed-integer linear program is developed to maximize the total profit of the value chain, involving decisions related to the volumes of wood to harvest and to keep in stock, as well as the quantities to deliver to each business unit to meet market demand. The model also includes different constraints concerning fibre freshness across the network (forest, sawmills, and paper mill), as this has an effect on operation costs and supply decisions. Various scenarios based on the demand variation, price fluctuations, and wood aging levels are tested, followed by a sensitivity analysis. Results show that price and demand variations as well as fibre freshness are important criteria to consider in the procurement and production planning for increasing the forest supply chain benefits. The scenarios analysed also confirm the usefulness of the model in guiding companies to make adequate planning decisions according to their business environment.

1. Introduction

The forest products industry is characterized by a complex logistics network that includes several stakeholders responsible for conducting many activities. In particular, it encompasses business units such as logging companies, sawmills, paper mills, transportation companies, wholesalers, retailers, etc. (Kaplinsky & Morris 2001). Improving supply chain planning typically involves optimizing material flows and developing effective operations management for value creation. However, in the forest sector, these challenges are very specific and they differ from traditional logistics networks (Martell et al. 1998). In particular, wood supply is a challenge due to the variation of fibre quality and variability in timber properties, induced by the heterogeneity of forest stands and the wide array of species composing some of them. Several other managerial elements can be challenging, such as harvest scheduling, transportation planning, and fibre freshness control. In such context, it becomes necessary to develop effective planning strategies to provide a product that meets quality expectations, ensures minimum delivery times, and generates minimum costs throughout the supply chain.

In this article, we describe a planning model based on a forest products supply chain. This type of supply chain is considered complex because it includes several independent business units responsible for conducting a relatively large number of interdependent activities (i.e. managing various forest stands, sawmilling operations, pulp and paper production, energy production, etc.).

Consequently, this work aims to develop a mathematical model that could help forest companies in planning and managing their procurement and production operations. The main motivation for modelling procurement activities is justified by the need to deal with wood fibre cost, quality, and availability. Planning a forest supply chain also requires decisions regarding harvesting time, amounts to be harvested, volumes to deliver to various customers, quantities to keep in stock at each node, capacities of each processing unit, and so on. Special attention is paid to wood fibre aging and its impact on the supply chain costs and profitability. The mathematical planning model proposed here is therefore different from previous work since it considers constraints for wood fibre freshness within a complex supply chain encompassing three levels: forest supply areas, sawmills and a paper mill, as well as the market. It also provides the information needed for companies to improve their decision-making process and achieve greater profitability through fibre quality tracking.

The paper is structured as follows: Section 2 presents a literature review. The description of the supply chain considered and the assumptions made for the mathematical modelling are then proposed in Section 3. Section 4 describes the mathematical model. Experiments are explained in detail in Section 5. Finally, Section 6 concludes the paper.

2. Literature review

Planning and control approaches in the field of supply chain management (SCM) aim at designing, implementing, and monitoring systems in order to improve business logistics networks as well as the interactions between network members (Martel 2003). Similarly, management of the forest supply chain involves key strategic decisions that may impact value creation and profit generation for all business units. Among the main foundations of forest SCM, we include the planning of logistics activities across several units such as forests and processing plants, as well as the flow connecting these units (Rönnqvist 2003).

In the literature, several planning approaches for supply chains have been developed to describe, design, and model forest-to-consumers operations. For example, Bredström et al. (2004), Gunnarsson and Rönnqvist (2008), and Lidestama and Rönnqvist (2011) have studied the problem of pulp supply for a large Swedish company. In the first study, the objective was to develop decision support models for logistics activities planning (i.e. production, transportation, and storage operations). In the two subsequent studies, the goal was to analyse the problem of integrated planning for the transportation, production, and distribution of pulp products to customers. The benefits from these studies are in their ability to produce realistic solutions in terms of production planning and flow optimization. Carlsson and Rönnqvist (2005) have also partnered with a forest company to improve and develop practical methods of management and optimization of the forest supply chain. The implementation of these methods provides a suitable environment for decision-making between the supply chain members. Chauhan et al. (2008) proposed an approach based on non-linear programming techniques as well as efficient heuristics for the problem of pulp and paper supply. The authors aimed to determine the best mix of parent rolls to keep in stock, and their allocation to processing plants, in order to minimize both trim losses and inventory costs.

Chauhan et al. (2009) developed a planning model that optimizes a forest supply chain in order to satisfy short-term timber demand for geographically distributed plants. Heuristic methods are used to find feasible solutions within a reasonable resolution time. On the other hand, the model does not take into account the quality of the product, more specifically, the selling price related to the quality offered and the processing cost associated with low quality fibres.

In fact, another particularity of the forest supply chain concerns fibre quality and freshness, as well as its effects on logistics activities at strategic, tactical, and operational levels. Favreau (2001) explained that several costs are affected by wood fibre aging, such as operational costs (e.g. the cost of chemical products purchased to improve the quality of pulp when chips are old), product yield, inventory cost (e.g. the inventory cost for unused materials at the wood yard), transportation cost, etc. He also showed how increasing storage duration may tend to decrease production yield because of the increasing dryness of the wood. As a result, research has been conducted based on these parameters. Beaudoin et al. (2007) developed a mathematical model for harvest planning through an efficient allocation of raw wood to mills and a reduction of processing and transportation costs. The model takes into account the freshness and the quality of wood associated with its age as well as the demand for logs, chips, and finished products. Maness and Norton (2002) developed a planning system which optimizes sawmill production, while considering the freshness of wood and the deterioration of fibre quality in mill inventories. The supply chain studied in these two articles was limited to two levels: forest supply areas and sawmills. Along the same line, Karlsson et al. (2003) proposed a model for forest operations planning. Their goal was to determine the harvesting schedules, the team assignment for harvesting activities, and the transportation and storage plans in the forest. Their model also takes into consideration the age and the quality of logs. More specifically, the authors associated storage costs to the age of logs stored in harvesting areas, terminals, and mills.

Other research fields address the key question of product freshness. For example, Ahumada and Villalobos (2009) offered a review of the major strategic, tactical, and operational planning models developed for the agri-food supply chain. In particular, they examined planning models aiming to maintain the quality and freshness of perishable products while describing the role of these models in strategic decision-making (e.g. network design, technology selection, profit maximization, costs reduction, etc.), tactical planning (e.g. harvesting planning, crop selection, labour capacity, etc.), and short-term decision-making (e.g. production planning, storage planning, packaging problems, etc.). As they target the objectives of improving the quality of finished products and reducing costs/maximizing profits, these models seem interesting for future modelling of natural resource supply chains. Nevertheless, the information given concerning the mathematical models and the formulation developed are insufficient to evaluate their potential adaptability for the forest industry. Similarly, Ahumada and Villalobos (2011) and Rong et al. (2011) proposed a planning model to manage both costs and product quality. Specifically,

Ahumada and Villalobos (2011) developed a model to manage crop maturity and yield while determining the effect of harvesting frequencies on the distribution of the fruit colour. Their study showed that an effective management of material freshness can lead to significant savings. On their side, Rong et al. (2011) developed a method for modelling food quality degradation and planning the food distribution system.

In the following sections, we study the case of a hypothetical logistics supply chain composed of independent forest companies. We propose a planning model that contributes to better synchronize operations, based on a centralized control approach. The model encompasses five forest supply areas, four sawmills, one paper mill, and two main customers, while assessing fibre freshness for wood, lumber, and chips.

3. Problem description

3.1. The forest products supply chain

Logistics activities in this type of supply chain start with forest operations, more specifically harvesting, transportation, and storage in the forest. Several other related activities are also required. We cite for example logging camp management, construction of haul roads, and allocation of teams to harvesting sites. Thereafter, a forest–sawmills procurement activity is part of the supply chain. For the purpose of this study, the forest products supply chain shown in Figure 1 was considered.

The forest shown in Figure 1 is the procurement area and consists of five sites. Each harvesting site can supply several sawmills with the same input or with different input. The variation of input mainly depends on quality criteria such as fibre freshness and tree diameters. Transportation between harvest sites and sawmills is ensured by log trucks managed on a weekly basis. Sawmills can also be supplied by external sources to cover the lack of supply in case of high demand (i.e. mainly auctions, private forests, and a few forest products companies). A bioenergy plant is also considered to reflect the increasing interest of forest products companies in creating such a factory for wood chips and forest

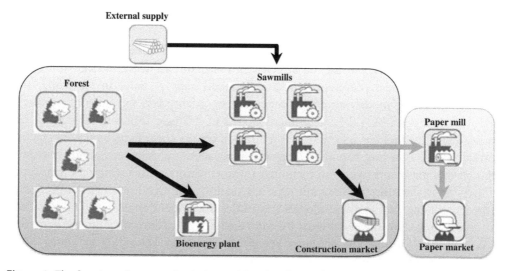

Figure 1. The forest products supply chain considered in this study.

residues. Supply for this facility is provided by available harvesting areas and the quantity to transport to the plant is modelled as a variable.

Timbers are transported to four sawmills with variable processing capacities. The wood required in terms of quantity, species, diameter, and quality may differ from one sawmill to another. Logs are consumed according to a first in first out rule. Sawmills consume logs to produce spruce and fir lumber as the main products, and spruce and fir chips (falls and waste) as a secondary product. Chips are sent to a single paper mill while lumber demand is generated by two customers. The paper mill uses chips to produce newsprint and magazine paper, in order to meet the demand for two clients. The type of products to produce, the quantities kept in stock at each node, the transported quantities, and the quantities produced by the processing units are among the decisions to make in order to satisfy demand.

Currently, the forest products industry faces many difficulties on several levels: reduced fibre supply, scarcity of labour, rising energy costs, seasonal demand, and high transportation costs. Moreover, while sawmills typically focus on maximizing value from lumber when planning their operations, with limited considerations to chip quality, paper mills have certain requirements regarding the chips provided by sawmills (i.e. in terms of quality and volume) to satisfy the paper demand. A chip quality that would not correspond to the needs would result in higher operation costs and lower paper sales price. On the other hand, paper mills are usually non-negligible customers of sawmills, and stopping their operation could sometimes lead to the closure of some sawmills as well.

In such a case, it is necessary to determine an efficient supply and control strategy that allows sawmills not only to respond effectively to lumber market demand, but also to support the competitiveness and profitability of paper mills by providing better chip quality. Demand satisfaction with well-defined quality and quantity criteria is complex, especially in a multi-functional context that gathers a large number of stakeholders. This is why we have developed a mixed-integer linear programming model to improve the planning process of the supply chain and ensure better use of wood fibre as well as efficient management of wood fibre freshness.

3.2. Modelling assumptions

The forest supply chain proposed in Figure 1 was modelled based on a few assumptions. First, we considered a one-year planning horizon divided into 52 weeks. Our experiment was performed on a four-week rolling horizon. Thus for each instance, the model is solved for the first four weeks, for example weeks 1–4, based on a known demand for the first period (i.e. week 1) and on a forecasted demand for the three other periods (weeks 2–4). The demand for a given week will be a forecasted demand until it becomes the first week of the rolling horizon. It then becomes a known demand. By using a rolling horizon, it is possible to combine both known and forecasted demands while taking into account updates, revisions, and potential corrections within the planning process. The approach is illustrated in Figure 2 for four periods.

In the planning model, we have considered a limited capacity for each harvesting area. Four types of raw material have been taken into account, specifically, small and large diameters of both fir and spruce logs. Regarding the demand for lumber and paper, it encompasses the type of product needed as well as the age expected (freshness). If a

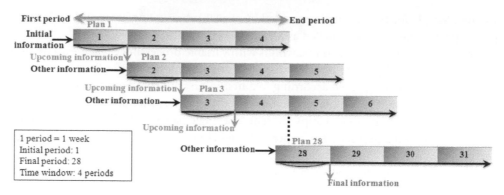

Figure 2. An example of the rolling horizon used in the experimentation.

supplier cannot satisfy the customer demand for a specific product quality, a better quality product is supplied instead at the same price as the undelivered product. The idea of penalizing the supplier by selling a product with higher quality at a lower price is thus part of a customer retention strategy. Each category of consumed logs generates a specific percentage of lumber and chips. We have also assumed that the paper mill uses a specific percentage of spruce and fir chips to produce each kind of paper, as observed in the case study. The following table summarizes the sets of products considered at each supply chain node and their units.

The level of freshness is divided into three categories. The first category is called green (young), the second one is called yellow (medium), and the third one is called red (old). We set θ as the percentage of aging at period t. The percentage of aging therefore means that when a given amount of product remains in stock during a certain period of time, the freshness level deteriorates. In our study, θ and the period t are invariable according to the product type and season. Because of this assumption, we also consider that $\varphi = \sqrt[t]{\theta}$, where φ represents the percentage of aging per time unit. If θ is, for example, a percentage of aging per week (7 days), the percentage of aging per day will be $\varphi = \sqrt[7]{\theta}$. Aging only a percentage of the stock is justified by the climatic and handling conditions in the lumber yard (as discussed by Ahumada and Villalobos (2011) and Rong et al. (2011)). Indeed, logs are usually stored in the lumber yard for a certain period of time before being

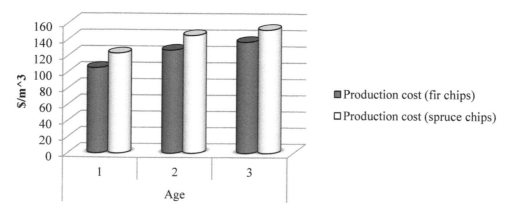

Figure 3. An example of the cost evolution for processing chips with regard to their age.

processed at the mills. They are grouped in stacks according to their physical characteristics (e.g. wood species, length, diameter, origin, etc.). As a result, from one period to another, not all the products will evolve in the same manner. Furthermore, production costs are defined according to the fibre age. Indeed, low quality fibre (old) requires a more complicated transformation process. For example, red chips require more chemical whitening to yield an adequate final product (paper). The following figure shows the cost evolution for processing chips depending on their age.

In addition, in order to ensure customer demand satisfaction in terms of quality and quantities, we added external sources encompassing, among others, public auctions and private forests to cover the lack of raw material when needed.

4. Mathematical modelling

In this section, we present the mathematical model developed to plan all of the activities of the proposed forest products supply chain (Figure 1). This model is based on a centralized control approach and aims to maximize the global network profit.

4.1. Modelling parameters and decision variables

Different sets and parameters have been used to formulate the planning model:

- T represents a set of periods.
- The forest products supply chain consists of several harvesting zones W.
- It is possible for these zones to deliver raw products to sawmills U and to bioenergy factories U'.
- The supply chain also includes an external supply source E whose mission is to satisfy sawmill orders in case of high demand or supply shortage from the harvesting zones (i.e. all of the different external sources are grouped into a unique node with an average supply cost).
- All sawmills U supply the paper mills U''. At the same time, they also satisfy the lumber demand for customers B.
- The paper mill delivers final products to several customers C. These products belong to the R class and are characterized by a level of freshness A.
- In the model, the A index represents the freshness categories, i.e. 1: green, 2: yellow, and 3: red.

An illustration of the modelling is presented in Figure 4.
The parameters used for the problem resolution are as follows:

C_{rwt}^{H}: Unit forest operations cost for product r for supply source w during period t ($\$/m^3$)

C_{reut}: Unit supply cost for product r from external supply source e to plant u during period t ($\$/m^3$)

C_{rwt}^{SW}: Unit storage cost for product r at supply source w during period t ($\$/m^3$)

C_{rwut}^{T}: Unit transporting cost for product r from source w to sawmill u during period t ($\$/m^3$)

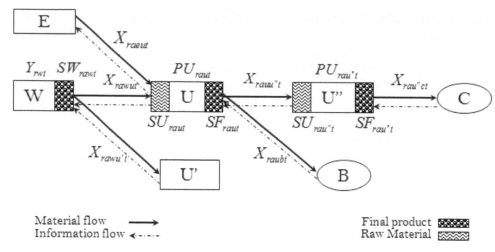

Figure 4. Illustration of the network studied.

$C^T_{rwu't}$: Unit transporting cost for product r from source w to plant u' during period t ($\$/m^3$)

$C^T_{ruu''t}$: Unit transporting cost for product r from sawmill u to paper mill u'' during period t ($\$/m^3$)

C^T_{rubt}: Unit transporting cost for product r from sawmill u to the lumber client b during period t ($\$/m^3$)

C^{SI}_{rut}: Unit storage cost for raw materials r at sawmill u during period t ($\$/m^3$)

$C^{SI}_{ru''t}$: Unit storage cost for raw materials r at paper mill u'' during period t ($\$/m^3$)

C^{SF}_{rut}: Unit storage cost for finished product r at sawmill u during period t ($\$/m^3$)

$C^{SF}_{ru''t}$: Unit storage cost for finished product r at paper mill u'' during period t ($\$/tonne$)

C^{PU}_{raut}: Unit production cost for product r, aged a, at sawmill u during period t ($\$/m^3$)

$C^{PU}_{rau''t}$: Unit production cost for product r, aged a, at paper mill u'' during period t ($\$/m^3$)

$P_{rau't}$: Selling price for product r, aged a, transported to plant u' during period t ($\$/m^3$)

P_{raut}: Price for finished product r, aged a, manufactured by sawmill u during period t ($\$/m^3$)

$P_{rau''t}$: Price for finished product r, aged a, produced by paper mill u'' during period t ($\$/tonne$)

α_{ru}: Conversion factor for converting raw materials into finished products r for sawmill u

$\beta_{ru''}$: Conversion factor for converting raw materials into finished products for paper mill u''

θ_{rwa}: Proportion of aging product r, aged a, for source w

θ^{si}_{rua}: Proportion of aging product r, aged a, for the initial stock at the sawmill u

θ^{sf}_{rua}: Proportion of aging product r, aged a, for the final stock at the sawmill u

$\theta^{si}_{ru''a}$: Proportion of aging product r, aged a, for the initial stock at the paper mill u''

$\theta^{sf}_{ru''a}$: Proportion of aging product r, aged a, for the final stock at the paper mill u''

b_{rwt}^{hmx}:   Maximum harvesting capacity of product r at source w during period t (m³)

b_{rwt}^{hmn}:   Minimum harvesting capacity of product r at source w during period t (m³)

b_{w}^{s}:   Maximum storage capacity of supply source w (m³)

b_{u}^{si}:   Maximum storage capacity for raw materials at sawmill u (m³)

$b_{u''}^{si}$:   Maximum storage capacity for raw materials at paper mill u'' (m³)

b_{u}^{sf}:   Maximum storage capacity of finished products at sawmill u (m³)

$b_{u''}^{sf}$:   Maximum storage capacity of finished products at paper mill u'' (tonne)

b_{tw}^{T}:   Maximum transportation capacity from source w during period t (m³)

b_{tw}^{TMin}:   Minimum transportation capacity from source w during period t (m³)

b_{te}^{A}:   Maximum supply capacity from external source e during period t (m³)

b_{tu}^{T}:   Maximum transportation capacity from sawmill u during period t (m³)

b_{ut}^{f}:   Maximum processing capacity at sawmill u during period t (m³)

$b_{u''t}^{f}$:   Maximum processing capacity at paper mill u'' during period t (tonne)

D_{raut}:   Demand for product r, aged a, of sawmill u during period t (m³)

$D_{rau''t}$:   Demand for product r, aged a, of paper mill u'' during period t (tonne)

M: A large number.

Two sets of decision variables have also been considered. The first one consists of the following binary variables:

$$ZU_{wt} = \begin{cases} 1, & \text{if wood is transported from the source } w \text{ to mill } u \text{ during period } t \\ 0, & \text{otherwise} \end{cases}$$

$$ZU'_{wt} = \begin{cases} 1, & \text{if here is a transport from the source } w \text{ to mill } u' \text{ during the period } t \\ 0, & \text{otherwise} \end{cases}$$

The second set of variables is defined as follows:

Y_{rwt}:   Harvested volume of product r from source w during period t (m³)

SW_{rawt}:   Volume of product r, aged a, stocked at supply source w during period t (m³)

X_{rawut}:   Transported volume of product r, aged a, from source w to sawmill u during period t (m³)

$X_{rawu't}$:   Transported volume of product r, aged a, from source w to plant u' during period t (m³)

X_{raeut}:   Transported volume of product r, aged a, from external source e to sawmill u during period t (m³)

$X_{rauu''t}$:   Transported volume of product r, aged a, from sawmill u to paper mill u'' during period t (tonne)

X_{raubt}:   Transported volume of product r, aged a, from sawmill u to lumber client b during period t (m³)

$X_{rau''ct}$:   Transported volume of product r, aged a, from paper mill u'' to final client c during period t (tonne)

SU_{raut}:   Inventory level of raw material r, aged a, at sawmill u during period t (m³)

SF_{raut}:   Volume of final product r, aged a, stocked at sawmill u during period t (m³)

$SU_{rau''t}$:   Inventory level of raw material r, aged a, at the paper mill u'' during period t (tonne)

$SF_{rau''t}$: Volume of final product r, aged a, stocked at the paper mill u'' during period t (tonne)

PU_{raut}: Inventory level of transformed product r, aged a, at the sawmill u during period t (m³)

PF_{raut}: Quantity of available finished product r, aged a, at the sawmill u during period t (m³)

$PF_{rau''t}$: Quantity of finished product r, aged a, at the paper mill u'' during period t (tonne)

$PU_{rau''t}$: Inventory level of transformed product r, aged a, at the paper mill u'' during period t (tonne)

4.2. Model formulation

The planning model developed maximizes the total profit of the supply chain proposed while satisfying customer demand for lumber and paper. The formulation of the objective function is defined as:

$$
\begin{aligned}
\text{Max} \quad & \sum_{r \in R} \sum_{a \in A} \sum_{u \in U} \sum_{b \in B} \sum_{t \in T} p_{raut} X_{raubt} + \sum_{r \in R} \sum_{a \in A} \sum_{w \in W} \sum_{u' \in U'} \sum_{t \in T} p_{rau't} X_{rawu't} \\
& + \sum_{r \in R} \sum_{a \in A} \sum_{u'' \in U''} \sum_{c \in C} \sum_{t \in T} p_{rau''t} X_{rau''ct} - \sum_{r \in R} \sum_{w \in W} \sum_{t \in T} C^H_{rwt} Y_{rwt} \\
& - \sum_{r \in R} \sum_{a \in A} \sum_{e \in E} \sum_{u \in U} \sum_{t \in T} C_{reut} X_{raeut} - \sum_{r \in R} \sum_{a \in A} \sum_{w \in W} \sum_{t \in T} C^{SW}_{rwt} SW_{rawt} \\
& - \sum_{r \in R} \sum_{a \in A} \sum_{w \in W} \sum_{u \in U} \sum_{t \in T} C^T_{rwut} X_{rawut} - \sum_{r \in R} \sum_{a \in A} \sum_{w \in W} \sum_{u' \in U'} \sum_{t \in T} C^T_{rwu't} X_{rawu't} \\
& - \sum_{r \in R} \sum_{a \in A} \sum_{u \in U} \sum_{u'' \in U''} \sum_{t \in T} C^T_{ruu''t} X_{rauu''t} - \sum_{r \in R} \sum_{a \in A} \sum_{u \in U} \sum_{b \in B} \sum_{t \in T} C^T_{rubt} X_{raubt} \\
& - \sum_{r \in R} \sum_{a \in A} \sum_{e \in E} \sum_{u \in U} \sum_{t \in T} C^{SI}_{rut} SU_{raut} - \sum_{r \in R} \sum_{a \in A} \sum_{e \in E} \sum_{u'' \in U''} \sum_{t \in T} C^{SI}_{ru''t} SU_{rau''t} \\
& - \sum_{r \in R} \sum_{a \in A} \sum_{u \in U} \sum_{t \in T} C^{PU}_{raut} PU_{raut} - \sum_{r \in R} \sum_{a \in A} \sum_{u'' \in U''} \sum_{t \in T} C^{PU}_{rau''t} PU_{rau''t} \\
& - \sum_{r \in R} \sum_{a \in A} \sum_{u \in U} \sum_{t \in T} C^{SF}_{rut} SF_{raut} - \sum_{r \in R} \sum_{a \in A} \sum_{u'' \in U''} \sum_{t \in T} C^{SF}_{ru''t} SF_{rau''t}
\end{aligned}
\tag{1}
$$

Thus, the objective function can be expressed as follows:

$$
Z = R^v - C^h - C^{ex} - C^t - C^s - C^m
\tag{2}
$$

where R^v is the revenue of the supply chain, C^h, the cost for forest operations, C^{ex}, the cost for buying wood from an external source, C^t, the total transportation cost (transportation between nodes: forest, plants, and customers), C^s, the inventory cost for the whole supply chain, and C^m, the cost for processing wood at the different business units.

Supply chain revenues are generated from sales of lumber and paper, and from the delivery of wood residues to the bioenergy plant. Costs are divided into several categories. Specifically, the forest operations cost includes the cost for harvesting operations, forest road construction and maintenance, as well as the administrative cost. The cost of external supply encompasses the costs induced by buying logs from external sources (purchasing cost, ordering cost, transportation cost, etc.). The transportation cost includes the product delivery cost as well as loading and unloading costs. There is also an inventory cost that includes, among other things, material handling and equipment costs. The processing cost then covers the costs and expenses for producing lumber, chips, and paper.

The problem constraints are defined by the following expressions:

$$\sum_{b \in B} X_{raubt} \leq D_{raut} \forall\ r \in R,\ u \in U,\quad t \in T,\ a \in A \tag{3}$$

$$\sum_{c \in C} X_{rau''ct} \leq D_{rau''t} \forall\ r \in R,\ u'' \in U'',\ t \in T,\ a \in A \tag{4}$$

$$PF_{raut} = \alpha_{ru} PU_{raut} \forall\ r \in R, a \in A,\ u \in U, t \in T \tag{5}$$

$$PF_{rau''t} = \beta_{ru''} PU_{rau''t} \forall\ r \in R, a \in A,\ u'' \in U'',\ t \in T \tag{6}$$

Forest areas

$$SW_{rawt} = \begin{cases} \theta_{rwa}\left[Y_{rwt} - \left(\sum_{u \in U} X_{rawut} + \sum_{u' \in U'} X_{rawu't}\right)\right] + (1 - \theta_{rwa})SW_{raw(t-1)} & (7) \\ \forall\ r \in R, a \in A, t \in T,\ w \in W, a = 1 \\ \theta_{rw(a-1)}\ SW_{r(a-1)w(t-1)} + (1 - \theta_{rwa})SW_{raw(t-1)} - \left(\sum_{u \in U} X_{rawut} + \sum_{u' \in U'} X_{rawu't}\right) \\ \forall\ r \in R, a \in A, t \in T,\ w \in W,\ a > 1 & (8) \end{cases}$$

Sawmills

$$SU_{raut} = \begin{cases} \theta^{si}_{rua}\left[\left(\sum_{w \in W} X_{rawut} + \sum_{e \in E} X_{raeut}\right) - PU_{raut}\right] + \left(1 - \theta^{si}_{rua}\right)SU_{rau(t-1)} & (9) \\ \forall\ r \in R, a \in A, t \in T,\ u \in U,\ a = 1 \\ \theta^{si}_{ru(a-1)}\ SU_{r(a-1)u(t-1)} + \left(1 - \theta^{si}_{rua}\right)SU_{rau(t-1)} + \left(\sum_{w \in W} X_{rawut} + \sum_{e \in E} X_{raeut}\right) - PU_{raut} \\ \forall\ r \in R, a \in A, t \in T,\ u \in U,\ a > 1 & (10) \end{cases}$$

$$SF_{raut} = \begin{cases} \theta^{sf}_{rua}\left[PF_{raut} - \left(\sum_{u'' \in U''} X_{rauu''t} + \sum_{b \in B} X_{raubt}\right)\right] + \left(1 - \theta^{sf}_{rua}\right)SF_{rau(t-1)} & (11) \\ \forall\ r \in R, a \in A, t \in T,\ u \in U,\ a = 1 \\ \theta^{sf}_{ru(a-1)}\ SF_{r(a-1)u(t-1)} + \left(1 - \theta^{sf}_{rua}\right)SF_{rau(t-1)} - \left(\sum_{u'' \in U''} X_{rauu''t} + \sum_{b \in B} X_{raubt}\right) \\ \forall\ r \in R, a \in A, t \in T,\ u \in U,\ a > 1 & (12) \end{cases}$$

Paper mill

$$SU_{rau''t} = \begin{cases} \theta^{si}_{ru''a}\left(\sum_{u \in U} X_{rauu''t} - PU_{rau''t}\right) + (1 - \theta^{si}_{ru''a})SU_{rau''(t-1)} & (13) \\ \forall\ r \in R, a \in A, t \in T,\ u'' \in U'',\ a = 1 \\ \theta^{si}_{ru''(a-1)} * SU_{r(a-1)u''(t-1)} + \left(1 - \theta^{si}_{ru''a}\right) * SU_{rau''(t-1)} - PU_{rau''t} \\ \forall\ r \in R, a \in A, t \in T,\ u'' \in U'',\ a > 1 & (14) \end{cases}$$

$$\mathrm{SF}_{rau''t} = \begin{cases} \theta^{\mathrm{sf}}_{ru''a}\left(PF_{rau''t} - \sum_{c \in C} X_{rau''ct}\right) + \left(1 - \theta^{\mathrm{sf}}_{ru''a}\right)\mathrm{SF}_{rau''(t-1)} & (15) \\ \forall\; r \in R,\, a \in A,\, t \in T,\; u'' \in U'',\; a = 1 \\ \theta^{\mathrm{sf}}_{ru''(a-1)}\,\mathrm{SF}_{r(a-1)u''(t-1)} + \left(1 - \theta^{\mathrm{sf}}_{ru''a}\right)\mathrm{SF}_{rau''(t-1)} - \sum_{c \in C} X_{rau''ct} \\ \forall\; r \in R,\, a \in A,\, t \in T,\; u'' \in U'',\; a > 1 & (16) \end{cases}$$

$$\sum_{r \in R}\sum_{a \in A} \mathrm{PU}_{raut} \le b^{f}_{ut}\forall\; u \in U,\; t \in T \tag{17}$$

$$\sum_{r \in R}\sum_{a \in A} \mathrm{PU}_{rau''t} \le b^{f}_{u''t}\forall\; u'' \in U'',\; t \in T \tag{18}$$

$$b^{hmn}_{rwt} \le Y_{rwt} \le b^{hmx}_{rwt} \;\forall t \in T,\; w \in W,\; \forall\; r \in R \tag{19}$$

$$\sum_{r \in R}\sum_{a \in A} \mathrm{SU}_{raut} \le b^{si}_{u}\forall\; t \in T,\; u \in U \tag{20}$$

$$\sum_{r \in R}\sum_{a \in A} \mathrm{SU}_{rau''t} \le b^{si}_{u''}\forall\; t \in T,\; u'' \in U'' \tag{21}$$

$$\sum_{r \in R}\sum_{a \in A} \mathrm{SF}_{raut} \le b^{sf}_{u}\forall\; t \in T,\; u \in U \tag{22}$$

$$\sum_{r \in R}\sum_{a \in A} \mathrm{SF}_{rau''t} \le b^{sf}_{u''}\forall\; t \in T,\; u'' \in U'' \tag{23}$$

$$\sum_{r \in R}\sum_{a \in A} \mathrm{SW}_{rawt} \le b^{s}_{w}\forall\; t \in T,\; w \in W \tag{24}$$

$$\sum_{r \in R}\sum_{a \in A}\sum_{u \in U} X_{raeut} \le b^{A}_{et}\;\forall\; t \in T,\; e \in E \tag{25}$$

$$\sum_{r \in R}\sum_{a \in A}\sum_{u \in U} X_{rawut} \le \mathrm{MZU}_{wt}\forall\; t \in T,\; w \in W \tag{26}$$

$$\sum_{r \in R}\sum_{a \in A}\sum_{u' \in U'} X_{rawu't} \le \mathrm{MZU}'_{wt}\forall\; t \in T,\; w \in W \tag{27}$$

$$b^{T\mathrm{Min}}_{tw}*\mathrm{ZU}_{wt} \le \sum_{r \in R}\sum_{a \in A}\sum_{u \in U} X_{rawut} \le b^{T}_{tw}\forall\; t \in T,\; w \in W \tag{28}$$

$$b^{T\mathrm{Min}}_{tw}\mathrm{ZU}'_{wt} \le \sum_{r \in R}\sum_{a \in A}\sum_{u' \in U'} X_{rawu't} \le b^{T}_{tw}\forall\; t \in T,\; w \in W \tag{29}$$

$$\sum_{r \in R}\sum_{a \in A}\left(\sum_{u'' \in U''} X_{rauu''t} + \sum_{b \in B} X_{raubt}\right) \le b^{T}_{tu}\forall\; t \in T,\; u \in U \tag{30}$$

$$\mathrm{SW}_{rawt}, Y_{rwt}, X_{rawut}, X_{rawu't}, X_{rauu''t},\; X_{raeut}, X_{raubt}, X_{rau''ct}, X_{rau''ct}, \mathrm{SU}_{raut}, \mathrm{SU}_{rau''t}, \tag{31}$$

$$\mathrm{PU}_{raut}\mathrm{PF}_{raut}\mathrm{PF}_{rau''t}\mathrm{PU}_{rau''t}, \mathrm{SF}_{raut}\mathrm{SF}_{rau''t} \ge 0 \forall (r, a,\; w,\; u,\; u',\; u'',\; t)$$

Equations (3) and (4) ensure customer demand satisfaction including the case where the demand concerns a product with an unavailable freshness quality (i.e. the mathematical model ensures demand satisfaction by delivering a better quality product at the same price as the one offered for the requested product). Equations (5) and (6) mean that the available quantity of the final product at sawmills and at the paper mill site must correspond to the quantity of final product produced and stocked. Parameters α and β refer to the conversion of raw material into final products. Equations (7) to (16) ensure flow conservation as well as aging for zones of storage at the forest, sawmills, and paper mill sites. More specifically, when a final product is at the processing plant, its age is 1. Next, the age of the item in stock or in transit increases over periods of time. In Equations (7), (11), and (15), the stock of the final products is a function of the volume produced and delivered as well as of the remaining amount of on-hand age (age 1). Equations (9) and (13) are the flow conservation constraints concerning the initial stock at each sawmill and at the paper mill. They reflect the stock degradation due to storage and handling delays. Besides, Equations (8), (10), and (14) define the initial stock aged more than 1 in storage zones. Equations (12) and (16) are used to calculate the stock of finished products aged more than 1 at sawmills and at the paper mill.

The equations linked to the processing capacity of each business unit are given in Equations (17) and (18). Equation (19) specifies the lowest and the highest harvesting capacities. The storage capacities of the storage zones are defined by constraints (20)–(24). Constraint (25) determines the maximum supply capacity from the external source. Equations (26) and (27) cover all possible cases where the transportation activity is triggered between supply sources and customers. Both minimum and maximum transportation capacities are defined by Equations (28) and (29). Equation (29) specifies the maximum carrying capacity between sawmills and their customers. The last constraint is used to ensure non-negativity.

5. Experimentation and discussion

In order to evaluate the profitability of the proposed forest products supply chain while assessing the planning model usefulness for different contexts, several scenarios have been conducted. We started by considering demand variability since it was reported as a major impediment to lumber production planning (Poirier 2004). A disturbance of market prices for lumber was also taken into account because the price for forest products has been known to vary greatly with time. Different levels of freshness were then considered in order to evaluate their effect on the supply chain profit. A higher aged wood fibre usually involves higher operational costs as pointed out by Favreau (2001). The generated instances were solved with CPLEX 12.4 by using a personal computer Intel Core i7-2600 CPU, 3.40 GHz with 16.00 GB of RAM. It took less than 60 minutes to solve each of the scenarios.

5.1. Variation in lumber demand

In North America, construction activities are mainly conducted during the summer because of climate conditions, leading to a variable demand for lumber with seasonal patterns. Based on this context, we decided to consider three different demand patterns, namely a fairly stable consumption (few perturbations), seasonality (fluctuations that recur annually during the same period), and a cyclical seasonality (cyclical fluctuations inside a seasonal demand). Figure 5 shows an example of the demand variation for spruce lumber.

The sets of considered products at each level are presented in the Table 1.

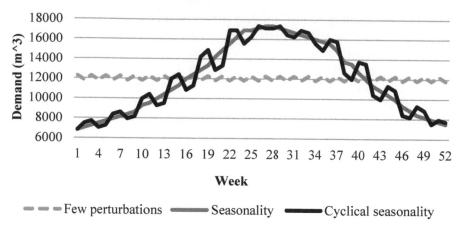

Figure 5. Three types of demand variation for spruce lumber, age 1 (1 m³ = 0.42 MBF).

The percentage of aging for this scenario was set to 79%, with lumber price varying from $138 per cubic meter to $179 per cubic meter, as presented in Table 2.

The results obtained from different variations of lumber demand are presented in Table 3.

It can be observed that higher supply chain profits are obtained when the demand exhibits seasonality (i.e. the highest) and cyclical seasonality. Since lumber prices are usually higher during high demand periods, the model seems to exploit this lucrative time of the year efficiently (see Table 4).

Table 1. Sets of considered products at each level.

Forest	Sawmills	Paper mill
Small and large size − fir logs (diameter and length) (m³)	Fir chips (m³)	Magazine paper (tonnes)
	Spruce chips (m³)	Newsprint (tonnes)
Small and large size − spruce logs (diameter and length) (m³)	Fir lumber (m³)	
	Spruce lumber (m³)	

Note: 1 m³ = 0.42 MBF (thousand board feet) (see http://cfs.nrcan.gc.ca/convcalcul?lang=en_CA).

Table 2. Lumber price and percentage of aging used for the scenario concerning the demand variability.

Period (weeks)	Lumber price ($/m³)	Percentage of aging φ (%)
1−17	138	79
18−34	179	79
35−52	160	79

Table 3. Results obtained for each instance considering demand variation for lumber in M$ (1 M$ = 1,000,000 $).

Costs Instances	Forest operations cost	Global transportation cost	External supply cost	Storage cost	Production cost	Total cost	Supply chain profit
Seasonality	63	25	8	1	31	128	5.8
Cyclical seasonality	62	25	8	1	29	125	5.1
Few perturbations	59	23	7	1	28	118	2.7

Table 4. Variations for each instance in lumber prices for spruce lumber, age 1.

Instance	Period (weeks)	Price ($/m³)	Average ($/m³)
S1	1–17	138	
	18–34	179	159
	35–52	160	
S2	1–17	179	
	18–34	160	159
	35–52	138	
S3	1–17	179	
	18–34	138	159
	35–52	160	

We can notice the importance of the forest operations cost and its variation depending on the demand pattern. This can be justified by the fact that forest companies have to harvest the maximum volume of wood allowed to satisfy their demand. Production and transportation costs are also more important when demand is characterized by a seasonality or a cyclical seasonality. It indicates an accumulation of stock during low demand periods and an increase of transportation and production costs during high demand seasons. The external supply cost seems, on the other hand, more constant and linked to the fact that when high quality materials are fully utilized, sawmills have to order from external sources to satisfy the demand with the right products.

Results confirm the model's ability to predict what would be the impact of different demand patterns on the supply chain profitability (i.e. higher demand, greater operation and logistics activities' intensity, lower demand, and smaller operation and logistics activities' intensity). The model can therefore become a useful tool to guide forest companies in planning their operations even in a complex setting.

5.2. Variation in lumber prices

Another important factor to take into account when modelling forest supply chain. This factor concerns the variation in lumber prices. Even though the price is typically set by the market, it becomes interesting to evaluate its impact on the supply chain profit. Therefore, three instances were tested based on a certain price variation. More precisely, we assumed three levels of price for each season (i.e. 17 weeks) distributed differently throughout the one-year planning horizon. The average price is kept the same. The percentage of aging for this scenario was set to 79%, with a seasonal demand.

The following table shows an example of price variation for spruce lumber, age 1.

Table 5 summarizes the results obtained when considering varying lumber price distributions, while Table 6 shows the profit variation per season.

Table 5. Results obtained for each instance when considering variations in lumber prices (in M$).

Costs Instances	Forest operations cost	Global transportation cost	External supply cost	Storage cost	Production cost	Total cost	Supply chain profit
1	63	25	8	1	31	128	5.8
2	61	24	9	1	31	126	4.1
3	62	24	9	1	31	127	1.7

Table 6. Comparison between generated supply chain profit per period for each instance.

Instance	Period (weeks)	Profit (in M$)	Total (in M$)
S1	1−17	−2.3	
	18−34	6.5	5.8
	35−52	1.6	
S2	1−17	1.2	
	18−34	3	4.1
	35−52	−0.1	
S3	1−17	1.2	
	18−34	−3.8	1.7
	35−52	0.9	

As expected, results show that the supply chain profit is sensitive to price disturbances. Indeed, for instances encompassing a higher price during a high season, the supply chain profit becomes much more interesting (Table 6).

Results confirm that the major impact market price variations may have on forest companies' profitability. While intuitive, such results point out the importance of anticipating even small price variations when planning supply chain activities in order to minimize their negative effect (as shown for example with Instance S3 during period 18−34).

5.3. Variation in wood fibre freshness

Wood freshness and fibre quality must be closely monitored to provide the required finished product quality to the customers. Nevertheless, most of the forest products companies do not take this element into account in their planning. For this reason, we decided to test the impact of wood fibre freshness on the profitability of the proposed supply chain. The instances considered are based on increasing percentages of aging, as shown in Table 7.

As an example, when $\varphi = 72\%$, it means that 72% of the remaining amount in stock aged a, during period t, is going to be aged $a + 1$ during period $t + 1$. The other 28% keeps the same age a, during period $t + 1$.

As highlighted in the literature, lower wood fibre freshness typically involves higher manufacturing costs such as higher chemical costs for producing the pulp. When building the scenario, we therefore took this fact into consideration. Specifically, we decided to reduce the production yield depending on the percentage of aging of the wood fibre. It involves that two different percentages of aging (e.g. $\varphi = 72\%$ and $\varphi = 87\%$) will not lead to the same output ratios.

As an example, when $\varphi = 72\% \Rightarrow \alpha = 34\%$ lumber and 66% chips; while $\varphi = 87\% \Rightarrow \alpha = 25\%$ lumber and 75% chips. On the other hand, each age category for this specific percentage of aging will be based on the same production yield. The same principle applies when producing a tonne of paper; different percentages of aging (e.g. 72% and 87%) will not involve the same chip ratio ($\varphi = 72\% \Rightarrow \beta = 75\%$ spruce chips and $\beta = 25\%$ fir chips; $\varphi = 87\% \Rightarrow \beta = 78\%$ spruce chips and $\beta = 22\%$ fir chips).

Table 7. Percentages of aging for each instance.

Instance	S1	S2	S3
Φ (%)	72	79	87

Table 8. Results obtained considering increasing percentages of aging for the wood fibre (in M$).

Costs Φ (%)	Forest operations cost	Global transportation cost	External supply cost	Inventory cost	Production cost	Total cost	Supply chain profit
72	64	26	8	1	30	129	6.8
79	63	26	8	1	31	129	5.8
87	61	24	7	1	31	124	4.2

The results found are summarized in Table 8.

It can be observed that when the aging percentage increases, the total supply chain profit decreases significantly, involving a reduction of 2.6 M$ between the minimum and the maximum aging rate. In fact, a higher aging percentage involves more material becoming older at each node of the supply chain from one period to another. As a result, it becomes even more difficult to meet the demand for high quality products and external supply becomes necessary to meet customer demand. Higher production costs also have to be incurred to generate high quality products. A reduction of the forest operations cost is associated to the fact that the material becoming old rapidly, a part of the volume needed will be supplied via external sources to avoid quality problems.

Figure 6 focuses on the variation of profit between the different instances when the aging percentage increases from one instance to another. This leads us to highlight the importance of quality management, including freshness, and proper handling and storage techniques. Since forest companies cannot prevent fibre from aging, they can, however, adopt appropriate planning and forecasting strategies for an effective control of their logistics activities.

5.4. Sensitivity analysis

In this section, we intend to determine how the variation of demand for lumber (D), different percentages of aging (φ), and lumber price (P) variation may affect the supply chain profit. We also want to determine whether these factors interact with each other.

We propose a 2^3 factorial plan (Law 2007) to calculate the effect of variables D (factor 1), φ (factor 2), and P (factor 3), on the total profit PT, PT $= f(D, \varphi, \text{ and } P)$.

The resulting three-factor factorial plan is described in Table 9. This method requires the choice of two levels for each factor, low and high. The numerical values are usually intuitively generated or determined after a thorough study of the case studied. In our case,

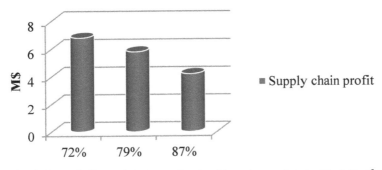

Figure 6. Supply chain profit for each instance (each with an increasing percentage of aging for the wood fibre).

Table 9. A 2^3 factorial plan and the factor values considered for low and high demands.

Factor	Low	High
D (1) (m³)	6829	3971
ϕ (2) (%)	87	72
P (3) ($/m³)	138	179

the values were selected after analysing data from governmental reports. The demand characterized by a seasonal pattern was used for the experimentation.

The values for demand presented in the previous table are limited to the demand for spruce lumber, age 1, during week 1. Figure 7 shows the variation pattern for low and high demand.

In a factorial plan 2^k, the number of estimated effects is equal to $2^k - 1$. In our case, we adopted a 2^3 factorial plan. Hence, the number of estimated effects is $2^3 - 1 = 7$. To compare the effect of the factors on each other, we assigned a minus sign to a lower numerical value and a plus sign to a higher numerical value. Table 10 shows the different combinations studied.

The evaluation of the effects related to the total profit is then as follows:

$$e_1 = \frac{-PT_1 + PT_2 - PT_3 + PT_4 - PT_5 + PT_6 - PT_7 + PT_8}{2^{k-1}}$$

We obtain: $e_1 = 2.66,$

$$e_2 = \frac{-PT_1 - PT_2 + PT_3 + PT_4 - PT_5 - PT_6 + PT_7 + PT_8}{2^{k-1}}$$

We obtain: $e_2 = -2.26,$

$$e_3 = \frac{-PT_1 - PT_2 - PT_3 - PT_4 + PT_5 + PT_6 + PT_7 + PT_8}{2^{k-1}}$$

We obtain: $e_3 = 7.37.$

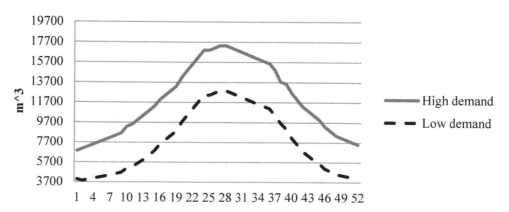

Figure 7. Weekly demand variation.

Table 10. Combinations used to measure the effects of three factors on the supply chain profit.

Point\effect	e1	e2	e3	e12	e13	e23	e123	Total profit (M$)
1	−	−	−	+	+	+	−	−5.8
2	+	−	−	−	−	+	+	−4.4
3	−	+	−	−	+	−	+	−3.7
4	+	+	−	+	−	−	−	0.8
5	−	−	+	+	−	−	+	3.6
6	+	−	+	−	+	−	−	11.4
7	−	+	+	−	−	+	−	9.3
8	+	+	+	+	+	+	+	14.2

We can thus estimate the interaction between the two factors e_{ij}. It can be measured using the difference between the average responses (total profit in our case) when both factors are simultaneously positive or negative. The interaction effects of two factors are completely symmetrical $e_{ij} = e_{ji}$, that is, $e_{12} = e_{21}$.

$$e_{12} = \frac{1}{2}\left[\frac{(PT_4 - PT_3) + (PT_8 - PT_7)}{2} - \frac{(PT_2 - PT_1) + (PT_6 - PT_5)}{2} \right]$$

Similarly, we can evaluate the interaction between three factors. In our case, this effect is equal to half of the difference between the means of the interaction effect of factors 1 and 2 when factor 3 is at its positive level and the means of the interaction effect of factors 1 and 2 when factor 3 is at its negative level.

$$e_{123} = \frac{1}{2}\left[\frac{(PT_8 - PT_7) - (PT_6 - PT_5)}{2} - \frac{(PT_4 - PT_3) - (PT_2 - PT_1)}{2} \right]$$

Results show that an increase of the demand by 2858 units per product per period decreases the profit by 2.66 M$ (Table 11), while an improvement of 41 $/m^3 of the selling price leads to a significant increase of the profit (7.37 M$). The aging effect is also considered as relatively important. More specifically, when the percentage of aging increases from 72% to 87%, the profit is reduced significantly (a decrease of 2.26 M$).

In addition, results show that the interaction between demand (1) and price (3), just as the average effect between the three factors (demand, freshness, and price), is relatively high. However, the effect between demand (1) and freshness (2), as well as between freshness (2) and price (3) is low. These results allow us to conclude that, for the forest supply chain proposed in this study, forest product prices are the main drivers of supply chain profit, followed by wood fibre freshness, and lumber demand. This analysis further confirms the importance of adjusting production plans according to forest product prices, while including the quality of the wood fibre in the decision-making process. The model proposed in this paper accounts for all of these factors.

Table 11. Values collected when analysing the effect of the three factors on the supply chain profit.

Effect	e_1	e_2	e_3	e_{12}	e_{13}	e_{23}	e_{123}
Value (M$)	2.66	−2.26	7.37	−0.05	2.4	−0.3	1.5

6. Conclusion

In this paper, we propose a mathematical model in order to plan harvesting, transportation, production, and storage operations for a supply chain composed of several forest products companies. The model was developed using a centralized control strategy that takes fibre freshness into account. Experiments were conducted based on three different scenarios reflecting the critical factors of the industry: demand variations, price variations, and aging of the wood fibre. Results show that wood freshness is an important criterion to consider in supply chain planning to adequately satisfy market needs.

Management of the forest products supply chain involves several critical issues such as customer satisfaction and synchronization between supply and demand. Based on these considerations, we propose a planning model that could help forest companies in delivering the right product to the right network node at the best moment.

Although the model developed gives interesting results, it is difficult to imagine implementing a fully centralized planning process even if it allows to better support the forest products supply chain. Each company composing the supply chain is independent and not necessarily willing to share private information with others. A model based on a coordinated planning strategy is therefore underway to take into account the autonomous planning of each business unit. By testing some collaborative approaches and different coordination mechanisms, we aim to generate a supply chain profit similar to the one obtained with the centralized model. Thereafter, we aim to propose a methodology that would support the implementation of the new approach to put forward. This requires identifying the players to involve, their respective roles and the information to share.

Disclosure statement

No potential conflict of interest was reported by the authors.

References

Ahumada O, Villalobos JR. 2009. Application of planning models in the agri-food supply chain: a review. Eur J Oper Res. 196:1−20.

Ahumada O, Villalobos JR. 2011. Operational model for planning the harvest and distribution of perishable agricultural products. Int J Prod Econ. 133:677−687.

Beaudoin D, LeBel L, Frayret J-M. 2007. Tactical supply chain planning in the forest products industry through optimization and scenario-based analysis. Can J For Res. 37:128−140.

Bredström D, Lundgren. J-T, Rönnqvist M, Carlsson D, Mason A. 2004. Supply chain optimization in the pulp mill industry-IP models, column generation and novel constraint branches. Eur J Oper Res. 156:2−22.

Carlsson D, Rönnqvist M. 2005. Supply chain management in forestry−case studies at Södra Cell AB. Eur J Oper Res. 163:589−616.

Chauhan SS, Frayret J-M, LeBel L. 2009. Multi-commodity supply network planning in the forest supply chain. Eur J Oper Res. 196:688−696.

Chauhan SS, Martel A, D'Amours S. 2008. Roll assortment optimization in a paper mill: an integer programming approach. Comput Oper Res. 35:614−627.

Favreau J. 2001. Identifying the cost impacts of wood storage using the Opti-Stock model. Advantage Restricted to FERIC Members and Partners. 2:1−8.

Gunnarsson H, Rönnqvist M. 2008. Solving a multi-period supply chain problem for a pulp company using heuristics−an application to Södra Cell AB. Int J Prod Econ. 116:75−94.

Kaplinsky R, Morris M. 2001. A handbook for value chain research. Ottawa: International Development Research Centre.

Karlsson J, Rönnqvist M, Bergström J. 2003. Short-term harvest planning including scheduling of harvest crews. Int Trans Oper Res. 10:413—431.

Lidestama H, Rönnqvist M. 2011. Use of Lagrangian decomposition in supply chain planning. Math Comput Model. 54:2428—2442.

Maness TC, Norton SE. 2002. Multiple period combined optimization approach to forest production planning. Scand J For Res. 17:46—471.

Martel A. 2003. Le pilotage des flux: concepts de base et approches contemporaines [Piloting flows: basic concepts and contemporary approaches]. Document in French. Document de formation [Training document], DF-3.1.1. La conception de réseaux logistiques [The design of Logistics networks]. CENTOR. Québec: Université Laval.

Martell DL, Gunn EA, Weintraub A. 1998. Forest management challenges for operational researchers. Eur J Oper Res. 104:1—17.

Poirier CC. 2004. Models for forecasting, demand management, and capacity planning. Using models to improve the supply chain. Boca Raton, FL: CRC Press, LLC. 87—112.

Rong A, Akkerman R, Grunow M. 2011. An optimization approach for managing fresh food quality throughout the supply chain. Int J Prod Econ. 131:421—429.

Rönnqvist M. 2003. Optimization in forestry. Math Program. 97:267—284.

Weinfurter S, Hansen EN. 1999. Softwood lumber quality requirements: examining the supplier/buyer perception gap. Wood Fiber Sci. 31:83—94.

Benefits of inter-firm relationships: application to the case of a five sawmills and one paper mill supply chain

Nadia Lehoux, Luc LeBel (iD) and Momen Elleuch

ABSTRACT

To access new markets and efficiently satisfy customer demand, many companies have decided to create collaborations with their suppliers, distributors, and even competitors. In this way, they can share the information needed to plan supply chain activities adequately while generating benefits that would not be achievable individually. In this project, we have studied the case of five sawmills that all shared the same important customer for their wood chip. That customer, a major paper mill, was facing significant difficulties and needed to benefit from more reliable, predictable and cost competitive wood flow from its suppliers. The main goal of the research was, therefore, to develop a collaboration scheme that would help these forest products companies to work together and to achieve in the long-term higher profitability. Results showed that sawmills could improve their profit by up to 44% by managing inventories for their main customer or jointly plan supply chain operations. Furthermore, based on game theory techniques such as Shapley value, we observed that greater collaboration may not be equally profitable for each supply chain member regarding their contribution. In our case study, the paper mill would need to share a portion of the benefits with its suppliers to ensure win—win collaboration.

Introduction

In today's economy, many companies have started working with key partners to facilitate information sharing and better synchronize supply chain activities. Even industries deemed 'traditional' such as the forest sector have started looking at collaborative approaches for planning and managing operations so as to deliver the right wood product to the right customer at a competitive price.

In Canada, the forest products industry is the most important employer of the country, providing up to 582,700 jobs. The industry's contribution is vital to many regions located far from city centres. In 2014, the value of Canadian forest product exports reached $30.8 billion (Natural Resources Canada, 2014). Nevertheless, during the last years, this industry has faced an increased international competition, rising operation and energy costs, reduced allowable cut on public land, and declining demand for paper, leading to a slowing economy and permanent closure of several companies. As a result, many forest products companies have shut down operations while other were forced to reconsider their way of doing business so as to reduce their operational costs while being more competitive on national and international markets.

In this project, we have studied the case of five sawmills (i.e. five competitors) and one paper mill all located in the Côte-Nord region in Quebec, Canada. In this supply chain, logs from public land are used by sawmills to produce different types of lumber. The manufacturing process also leads to the production of wood chips as by-products. Once delivered to the paper mill, wood chips are mixed with chemicals and water to produce pulp and paper. Even though the paper mill purchase of wood chips is critical for sawmills profitability, each sawmill made planning decisions by focusing largely, if not solely, on sawmilling operations, leading to poor wood chip quality. Because no structured collaboration scheme was applied, each sawmill planned its operations in order to minimize its own costs rather than the global cost of the supply chain. For this context, we have proposed to develop a collaboration scheme that would help these stakeholders to work together while ensuring in the long-term higher profitability for the industry. This collaboration scheme relies on optimization techniques as well as game theory approaches. As a result, we have first conducted interviews with companies' managers in order to understand their current procedures while capturing their expectations regarding collaboration. Next, we selected three logistics strategies that could contribute to improve the coordination of sawmills and paper mill operations as shown in past research: Regular Replenishment (RR), vendor managed inventory (VMI), and Collaborative Planning, Forecasting and Replenishment (CPFR) (see for example the work of Lehoux et al. 2014). Using real data from the companies as well an adaptation of a software platform called Logilab (Jerbi et al. 2012), we have then evaluated what benefits these strategies could yield for five different contexts. The first scenario only considered three sawmills because at the moment of the study, two forest products companies decided to temporarily close their business units located in the region. The second scenario analysed the impact of using an external supplier (i.e. located in a different Quebec region) to supply wood chips to the paper mill. The third scenario took into account the possibility of sorting wood species for delivering improved chip quality. The fourth scenario involved managing chips freshness on a first-in, first-out basis. The last scenario supposed 'an ideal context' where all five sawmills would operate and supply wood chips to the paper mill. Results showed that sawmills could improve profit by up to 44% by implementing collaborative techniques such as VMI or CPFR. Similarly, by changing their way of sorting wood species or managing wood fibre freshness, supply chain profit could increase by up to 31%.

Techniques from game theory were also put to contribution (i.e. Shapley value, separable and non-separable costs, and a technique adapted from Kwasnica et al. 2005) to make sure that such collaboration scheme would be profitable for both the sawmills and the paper mill. We observed that one sawmill should achieve a higher profit than the one it gets presently, because the high-quality wood chips it supplies to the paper mill contributes greatly to the supply chain profit improvement. Furthermore, game theory techniques demonstrated that advanced logistic strategies are less profitable for sawmills, and consequently the paper mill would need to share a portion of its supply chain profits to ensure win—win relationship. The adaption of Kwasnica et al.'s (2005) price allocation technique in combinatorial auctions for evaluating individual contributions gave similar results as the individual profits obtained from the optimization. To the best of our knowledge, it is the first time that such a technique is adapted to calculate collaboration profit sharing.

Based on this case study, we may observe that a better coordination between sawmills' operations and paper mill activities could be profitable individually and collectively. If forest products companies of the study region would be willing to share information, manage wood chips for the paper mill or jointly plan supply chain operations, less wood chip deterioration would be generated and higher quality products would be produced. The case study also pointed out the fact that greater collaboration is not necessarily equally profitable for all the supply chain members, involving the necessity for implementing a strategy that ensures fair profit sharing.

The paper is structured as follows: we first describe the concept of inter-firm collaborations through a literature review. Next we introduce the case study, the methodology followed to conduct the analysis, the mathematical modelling and the results obtained from the experiments. Finally, a look at the implementation challenges and some concluding remarks regarding the merit and limitations of our approach are provided.

Literature review

Inter-firm companies can refer to two or more autonomous companies that work together in order to jointly plan and manage operations while generating benefits that could not be obtained individually (Stadtler 2009). Different types of collaboration schemes can, therefore, be put into practice depending on the objectives pursued, but authors usually classify them into three categories: horizontal collaborations, vertical collaborations and lateral or synergistic collaborations (Audy et al. 2012). When two or more companies belonging to the same supply chain work together in order to synchronize their operations and share more information, we call it vertical relationship. When two companies from different supply chains such as two competitors decide to share their transportation capacity or warehousing space, it is thus an example of horizontal relationship. We sometimes find both types of collaboration, two or more companies in the same supply chain that work together while collaborating with another group of companies from another supply chain. It is, therefore, called lateral collaboration (Mason et al. 2007). This form of collaboration is the one related to our study.

When companies are willing to build trust and allocate enough resources to support the relationship, collaboration benefits may be significant (Cao et al. 2010). Lehoux et al. (2008), Camarinha-Matos et al. (2009), Ramanathan and Gunasekaran (2014) and MacCarthy and Jayathne (2012) have all identified many advantages in implementing inter-firm collaborations: competencies and technologies sharing, access to new markets, costs reduction, increased supply chain visibility, better implementation of sustainable practices, etc. Nevertheless, because inter-firm collaborations may also involve important changes in the organization and even some risk, companies should take the time to develop and manage them carefully (Fabbe-Costes & Lancini 2009). According to Lehoux et al. (2014), establishing collaboration involves four main steps.

The first step requires building the relationship by selecting a partner and establishing the legal framework. The second step necessitates the implementation of coordination mechanisms to support planning decisions and improve overall benefits. A first key mechanism concerns information sharing to improve supply chain visibility and facilitate the decision-making process (Simatupang & Sridharan 2008). Logistics strategies can also be put into practice to better synchronize partners' activities. For example, with VMI, the manufacturer manages the inventories of its products for the buyer. This strategy can be used to improve both the replenishment process and production planning (Marquès et al. 2010). Collaborative Planning Forecasting and Replenishment is another technique used to improve forecasting accuracy and network planning (Büyüközkan & Vardaloğlu 2012). The third step concerns the collaboration outcome. Performance must be measured to make sure that the relationship is profitable for everyone, based on tools and indicators defined during the first step. Techniques from game theory can be useful to evaluate the contribution of each stakeholder. For example, coalition gains can be shared based on a fair distribution between partners (Shapley value) or profit divided into two parts, separable and non-separable profit (separable and non-separable costs) (Frisk et al. 2010). The final step involves implementing some incentives to share collaboration benefits fairly between partners, eliminate opportunistic behaviours and keep individual strategies aligned with the collaboration objectives. Price agreement, revenue sharing and quantity discounts are examples of incentives frequently used by companies (Cachon 2003). In our study, we focus on step two and three, by identifying adequate logistics strategies for the partnership while measuring their profitability for both the producers (i.e. sawmills) and the main customer (i.e. paper mill).

Several reasons make inter-firm collaborations particularly interesting for the forest industry. First, several companies are usually involved in supplying, manufacturing and delivering forest products, each being responsible for one or multiple operations along the supply chain. As a result, planning these business processes always involves managing relationships across complex networks of companies. Moreover, all activities are linked, and the more these operations need to be synchronized, the more accurate and timely the information to share must be. Lefaix-Durand et al. (2006) pointed out the necessity for forest products companies to create partnerships in order to react more rapidly when any event or change in the business environment occurs. Beaudoin et al. (2010) explored collaboration opportunities in developing harvesting plans for companies that get the wood fibre from the same public forest areas. Palander and Vaatainen (2005), Audy et al. (2007) and Frisk et al. (2010) studied collaboration for wood transportation as well as different techniques that could be used to redistribute the benefits between partners. Lehoux et al. (2011) studied different coordination mechanisms to facilitate collaboration between a pulp and paper company and its wholesaler, while Chambost et al. (2009) considered the creation of partnerships as a way for facilitating forest biorefinery implementation. Naud and D'amours (2011) suggested the establishment of inter-firm collaborations for ensuring the development of a sustainable forest supply chain. The previously mentioned research yielded useful insights concerning methodologies and benefits achievable through collaboration. Nevertheless, no application addresses the complex common case of the divergent processes associated with the interconnected operations of sawmilling and paper producing, despite the fact that this combination is the most common supply chain organization in the forestry sector. In the following sections, we will, therefore, provide evidence that collaboration could be beneficial for this key channel, based on different logistics strategies.

Description of the case study

The case studied in this research concerns different business units located in the Côte-Nord region in Quebec, Canada. This remote area represents more than 20% of the surface of the Province. Over the last five years, the global economic slowdown had a dramatic impact on the region, leading to a loss of 6300 jobs. The forest industry is now the second most important employer for the region, behind mining, whereas it was the first one a couple of years ago. The government has also reduced allowable cut on public land by 32% for the 2008–2013 period, leading many companies to excess production capacity.

The region has only one paper mill to which five different sawmills send most of their wood chips production. Of the five sawmills operating in the region, one is owned by the same company operating the papermill (Figure 1). While all five sawmills sell lumber for the construction market, it was reported to us that wood chips had to be sold for ensuring profitable operations. Consequently, closing the paper mill would likely mean the end of lumber production in this region. In other words, no entity had complete control over the supply chain or could operate independently of the other units. The paper mill depended on high-quality wood chips to produce paper meeting its clients requirements and be profitable, while sawmills had to sell their wood chips as well as their lumber to reach profitability. In theory, these companies could improve their individual and cluster performance by coordinating operations.

Nevertheless, at the moment of the study, collaboration was limited. The paper mill was the main, when not the only chip customer for the sawmills, yet sawmills' planning decisions were made focusing strictly on lumber quality, largely neglecting chip quality. Each sawmill procured logs from different forest areas. Ratio of spruce and fir harvested as well as wood quality differed from one sawmill to another. In addition, each sawmill had its own approach for sorting wood and producing the chip. Species were usually mixed at different stages of the production process so it became difficult to predict the quality of the final wood chip. Paper quality is directly linked to chip quality in terms of brightness, freshness and density. Species and age are also key factors to control. Lower chip quality, or a mismatch between quality standard and a specific paper product, leads to higher production cost or missed opportunities for the paper mill. Once chips have become too old, they cannot be used in paper production, so they have to be burnt or simply thrown away. When independent sawmills could not satisfy the demand of the paper

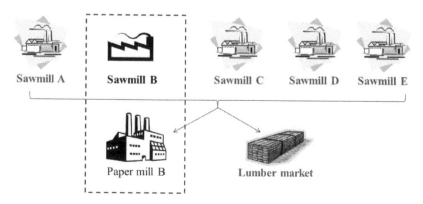

Figure 1. Illustration of the case study.

mill in terms of volume or quality, sawmill B, which is owned by the papermill, has to fill the gap. This can be done by chipping whole trees which usually leads to lower density chips. Based on this context, we have tried to evaluate how forest products companies in this region could collaborate so as to:

- better use the region's wood supply potential via inventory minimization and wood chips precision deliveries;
- respond more efficiently to the paper mill demand in terms of chip quality and volume; and
- provide profit opportunities for both sawmills and the paper mill.

Methodology

The case study was conducted following a four-stage approach (Figure 2). We have first tried to understand the current relationship between the sawmills and the paper mill via different interviews with people from the companies and in-situ observations during field trips.

After noticing poor coordination of the supply chain activities, we decided to evaluate from a theoretical point of view whether the use of three recognized logistics strategies could help in improving wood chips and paper quality while having a positive impact on the profit of each player. The strategies selected were: 1) RR, 2) VMI and 3) CPFR. With RR, each sawmill has to deliver a specific quantity of chips on a regular basis (e.g. X tonnes per week), based on the paper mill needs. If the strategy implemented is VMI, sawmills are responsible for managing chip inventory for the paper mill. The stock level at the paper mill is kept between a minimum and a maximum level to guarantee a certain service level. This kind of strategy involves real collaboration between sawmills (i.e. horizontal collaboration) to share information and make sure that paper mill needs are continuously satisfied. As a result, we have considered a coalition of sawmills rather than five autonomous decision-making units for this logistics strategy. With CPFR, partners jointly plan

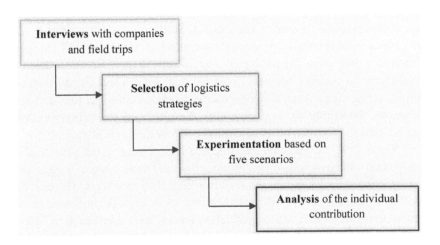

Figure 2. Methodology followed for the case study.

Table 1. Scenarios used to test each logistics strategy.

Scenarios	Description
1. As-is operations	--→ Three sawmills supplying wood chips
	--→ Species not sorted
	--→ No management of the chip freshness
2. External supplier	--→ Chips bought by the paper mill from an external supplier
	--→ High quality chips at a higher price
3. Wood species sorted	--→ Species sorted at the sawmill sites
	--→ Two categories, spruce chips and fir chips
4. Management of the chip freshness	--→ Management of chip freshness based on a first-in, first-out basis
	--→ Three categories used:
	Fresh = 0–5 days
	Less fresh: 6–10 days
	Not fresh: 11 days and more
5. Five sawmills	--→ Wood chips supplied by five sawmills rather than three

operations based on a common forecast. It is, therefore, modelled as an integrated planning model.

In order to estimate their usefulness in better coordinating the supply chain activities, each logistics strategy was tested based on five different contexts (Table 1).

(1) The first scenario reflected the sawmills' current way of doing. In particular, three sawmills were able to supply wood chips to the paper mill, as two forest products companies had temporarily closed their business units located in the region at the moment of the study. Furthermore, wood species were not sorted at the sawmill sites and wood chip freshness was not managed.

(2) In Scenario 2, the paper mill had the possibility to buy high quality chips from an external supplier at a price higher than the one charged by the three operating sawmills. From a practical point of view, this supplier is located far from the paper mill (i.e. in another Quebec region) and transportation cost is significantly higher. This supply is also uncertain, meaning that an order to this supplier may stay unsatisfied. Thus, the idea of modelling a second supply source was to evaluate the impact of higher wood chip quality supply in the process rather than considering a potential partnership with this external player.

(3) In Scenario 3, we supposed that chips were sorted based on wood species. In this way, it was possible to deliver the exact quantity needed by species (i.e. $x\%$ of fir and $(100-x)\%$ of spruce), easing process control at the paper mill. At the moment of the study, one sawmill had already started sorting its wood chips. Even though it was difficult to evaluate the exact cost for managing wood chips inventory differently all along the process, we knew that it was achievable and that it could have an impact on the supply chain profitability. Results from the experimentation could then be used to convince other sawmills to follow the same steps.

(4) With Scenario 4, we analysed the impact of managing wood fibre freshness on a first-in, first-out basis. More specifically, three different categories were used: wood chips produced and kept in stock during less than five days (i.e. one production week) were considered as fresh; wood chips kept in stock during more than five days but less than ten days (i.e. two production weeks) were considered as less fresh; and wood chips kept in stock more than 11 days were considered as not fresh anymore. The fresher the chip used in the process, the lower the paper production cost.

(5) Finally, in Scenario 5, we assumed that all five sawmills supplied chips to the paper mill. The probability that the two sawmills temporarily closed cease their production permanently was very high. On the other hand, we aimed to demonstrate the significant impact on the region economy if all sawmills would be able to supply the paper mill adequately.

Experiments were conducted using Logilab, a software platform developed by the Forac research consortium to model and optimize supply chains from the forest to the final customer (Jerbi et al. 2012). The user can insert business units such as sawmills, warehouses and paper mills on a map, while specifying inputs, outputs and processes of each site (Figure 3). Logilab will next evaluate the distance between each business unit and then optimize the supply chain cost (transportation, production, and inventory cost) or the supply chain profit (revenues from chips, lumber, paper, and energy) depending on the choice of the user. An optimization model (i.e. a linear programming model) is used to identify the quantity to produce, to keep in stock and to deliver to each business unit for each product defined (the mathematical model is presented in detail in Jerbi et al. 2012).

The optimization model ran by Logilab had to be adapted to account for specific dimensions of the problem. First, in order to measure the profitability of VMI for our case study, a new functionality in the software was added to track the inventory level at the paper mill site and make sure that it was kept between a minimum and a maximum level. Next, rules concerning the management of chip freshness were added in order to test the fourth scenario.

Figure 3. Illustration of the case study modelled using Logilab.

Higher coordination of the supply chain operations and an adequate supply of wood chip quality can certainly lead to cost reductions (e.g. lower paper production cost, lower inventory level, wood chip waste minimization, etc.). Nevertheless, a question remains: will this improvement be profitable for everyone? No company will usually be willing to collaborate if it involves losing money. As a result, we investigated how profit obtained from greater supply chain coordination should be shared between partners. In particular, we compared the portion of profit each company should get regarding its contribution to the collaboration (i.e. a coalition) with the one obtained from the supply chain optimization. The grand coalition considered was, therefore, composed of four players: three sawmills and the paper mill (i.e. the analysis was not conducted for Scenario 5 with five sawmills). This aspect of our research project was capital since a clear protocol for benefit sharing has been identified as the main obstacle to the implementation of collaborative supply chain coordination (Audy et al. 2012).

Three techniques were used to compare different potential solutions, the Shapley value representing the expected marginal contribution of a player when it enters into the coalition (Myerson 1997), separable and non-separable costs where the separable profit is allocate to each player while the non-separable part is distributed among the players based on certain weights (Frisk et al. 2010), and a technique adapted from an auction method proposed by Kwasnica et al. (2005). In this approach, the idea is to use a linear programming model to evaluate the profit to allocate to each player in order to satisfy core conditions while guaranteeing a certain uniformity regarding profit sharing between the different coalitions. The reason for using three different techniques was not to identify the best one but rather to compare different possible profit sharing solutions for the case study. This analysis could then be used to guide companies in defining a partnership agreement among coalition members. The mathematical models for estimating the benefits to share as well as the results obtained are summarized in the following sections.

Mathematical modelling

Mathematical models based on game theory techniques were used to determine the part of the collaboration profit that should be allocated to each player regarding its contribution to the coalition. We only summarize here the key mathematical expressions used for the case study but the reader can look at the work of Boyer et al. (2006) and Frisk et al. (2010) for more details.

Consider a coalition of S players as a subset of a grand coalition N. For the case study, there were three open sawmills (S) that wanted to collaborate with one paper mill (P). The grand coalition can, therefore, be composed of four players, namely $N = \{S1, S2, S3, P\}$. $P(S)$ is the profit generated from a subset S of participants and $P(N)$ the profit of the grand coalition. The key goal of the forest products companies was to improve the overall supply chain profit. If we denote y_i, the part of the global profit obtained by the participant i, the total amount received by the participants could be:

$$\sum_{i \in N} y_i = P(N) \tag{1}$$

The profit sharing could thus be considered as group rational or efficient (1). They also wanted to make sure that the profit obtained from the collaboration would not be lower than the one generated if they would have played alone.

$$\sum_{i \in S} y_i \geq P(S), \forall S \subseteq N \tag{2}$$

In this case, the profit sharing could be qualified as individually rational or stable (2). All players agreed that if the two temporarily closed sawmills would join the coalition, the impact on the overall supply chain profit would be positive:

$$\forall S, T \subseteq N : P(S \cup T) \leq P(S) + P(T) \tag{3}$$

The profit sharing could, therefore, be considered as sub-additive (3).

Focusing on the need to improve the economic situation of the region, we decided to use techniques that would contribute to minimize coalition secessions (Boyer et al. 2006). The first technique used concerns the Shapley value, a method that satisfies the group rational propriety but not necessarily the individually rational one (Frisk et al. 2010). With this approach, y_i can be measured using the following expression:

$$y_i = \sum_{S \subseteq N: i \in S} \frac{(|N| - |S|)!(|S| - 1)!}{|N|!} [P(S) - P(S - \{i\})] \tag{4}$$

Another method concerns separable and non-separable costs or, in our case, separable and non-separable profits. The idea is to first allocate the separable profit among the participants. Next, the non-separable part may be distributed equally (i.e. ECM or equal charge method) or based on a certain weight w_i that reflects the benefits participants will get when joining the grand coalition (i.e. ACAM or alternative cost avoided method). ACAM was the one used in our study. With this technique, y_i can be calculated using the following expression:

$$y_i = m_i + \frac{w_i}{\sum_{i \in N} w_i} g(N) \qquad \forall i \in N \tag{5}$$

The separable profit m_i and the non-separable profit $g(N)$ are evaluated using:

$$g(N) = P(N) - \sum_{i \in N} m_i; \; w_i = m_i - P_i; \; m_i = P(N) - P(N - \{i\}) \qquad \forall i \in N \tag{6}$$

The group rational propriety can again be satisfied. A third method is based on the work of Kwasnica et al. (2005). These authors proposed a mechanism to evaluate individual prices for combinatorial auctions. In particular, this type of auctions allows bids on packages of discrete items. Once the winning bid has been determined, it becomes necessary to evaluate the individual price for all the items included in the combinatorial bid. The sum of these individual prices should be equal to the winning bid as well as higher than all the losing bids. This approach can, therefore, reflect our context. The profit of the grand coalition needs to encompass all the individual contributions of its members while ensuring that it is profitable to be part of the grand coalition.

Based on their formulation, we have developed a linear programming model that evaluates the part of the profit each player may obtain from the grand coalition while ensuring uniformity when sharing profit between the different coalitions. To the best of our knowledge, it is the first time that such a technique is adapted to calculate collaboration profit sharing. In particular, the model will try to share profit so as to uniformly maximize the difference between $\sum_{i \in S} y_i$ and $P(S)$ for each coalition of N as follows:

$$\text{Max } z \tag{7}$$
$$\text{subject to}$$

$$\sum_{i \in S} y_i - P(S) \geq g_S \quad \forall \, S \subset N \tag{8}$$

$$\sum_{i \in N} y_i = P(N) \tag{9}$$

$$g_S \geq z \quad \forall \, S \subset N \tag{10}$$

$$y_i, g_S, z \geq 0 \tag{11}$$

The model contributes to maximize $\min\{\sum_{i \in S} y_i - P(S), S \subset N\}$ while taking into account the core conditions. This mechanism is similar to the nucleolus allocation method which aims to minimize the maximum excess or unhappiness of each coalition lexicographically (Okada et al. 2009). However here g_s is used so as to balance the prices (i.e. the part of the profit obtained by each member) across the items (i.e. the coalition members) (see Iftekhar et al. (2009) for an exhaustive comparison of the two methods).

The following section will show how these mathematical models have been used to support profit sharing for our case study.

Experimentation, results and analysis

In order to evaluate how the five sawmills and the paper mill could work together to better coordinate their operations and improve their profit, different experimentations were conducted. All the data used were based on the case study and collected from both the companies and technical reports. We used a planning horizon of one week for a total planning period of one year to conduct the experimentation. Each business site had a production capacity and we assumed the same timber supply cost for the five sawmills (i.e., the individual cost was unknown). Because the Côte-Nord's forest is mainly made up of fir and spruce, two types of chips were taken into account.

First scenario

We have first compared RR, VMI and CPFR for the current context, which is three sawmills supplying the paper mill without any specific sorting rules. As anticipated, results

Table 2. Individual and supply chain profits obtained for Scenario 1.

| | Scenario 1 | | |
	RR	VMI	CPFR
Sawmill A	$3,397,886	$4,816,104	$4,798,046
Sawmill B	$2,643,240	$-889,114	$-889,235
Sawmill C	$6,695,885	$7,715,799	$7,717,897
Paper mill	$21,663,168	$26,934,317	$27,654,251
Supply chain	$34,400,179	$38,577,106	$39,280,960

showed that the more companies are willing to collaborate, the better individual and supply chain profits could be (Table 2).

In particular, except for sawmill B, each partner gets a higher profit when sawmills manage chips inventory at the paper mill site or when operations are jointly planned based on common forecast. Because only three sawmills can deliver chips to the paper mill and since no external supplier is considered, sawmill B is often required to chip whole trees. In this way, the demand of the paper mill can be satisfied and its profit improved. On the other hand, sawmill B's demand for lumber stays unsatisfied, resulting in a significant lost in profit. In fact, this sawmill is the one located nearest to the paper mill and is owned by the same company as the paper mill. Therefore, if we look at the overall profit of sawmill B, we observe that losses are reduced by up to 10% when logistics strategies such as VMI or CPFR are implemented, even though the profit of sawmill B is negative.

Theoretical results reflect the strategy applied by company B at the moment of our study. Sawmill B was exclusively used to produce wood chips (i.e., chip whole trees) for the paper mill. The market price for lumber was relatively low and company B thought that it would be more profitable to focus on paper demand satisfaction. This situation was on the other hand not considered sustainable by the company. Instead, this lead decision-makers to reconsider the potential contribution to the network of the two temporarily closed sawmills.

Scenario 2

In Scenario 2, the paper mill has the possibility to buy high-quality wood chips from an external supplier at a higher price than the one charged by the three sawmills. As shown in Table 3, the paper mill will use the external supplier to fill its supply gap. Sawmill B will, therefore, stop chipping whole trees and its operating balance will be positive (profit). Supply chain profit and individual profit are still higher when advanced logistics approaches are used.

The external supplier is located very far from the paper mill so the transportation cost for wood chips is significant. This supplier may also refuse to supply wood chips at each period, prioritizing its local market. We, therefore, consider this external player as a

Table 3. Individual and supply chain profits obtained for Scenario 2.

| | Scenario 2 | | |
	RR	VMI	CPFR
Sawmill A	$3,397,906	$4,880,112	$4,879,092
Sawmill B	$2,631,224	$3,271,196	$3,280,181
Sawmill C	$6,717,766	$7,712,215	$7,712,723
Paper mill	$26,547,500	$26,818,336	$26,826,989
Supply chain	$39,294,395	$42,681,859	$42,698,985

Table 4. Individual and supply chain profits obtained for Scenario 3.

	Scenario 3		
	RR	VMI	CPFR
Sawmill A	$3,370,010	$4,875,635	$4,879,121
Sawmill B	$2,643,985	$3,219,448	$3,276,271
Sawmill C	$6,700,570	$7,712,966	$7,711,747
Paper mill	$31,020,689	$31,542,257	$32,125,866
Supply chain	$43,735,255	$47,350,306	$47,993,005

punctual solution to increase wood chip supply, but it would be difficult to see it as a strategic partner due to its geographic location.

Scenario 3

It is now assumed that wood species are sorted at the sawmill sites. As a result, sawmills can deliver species-specific chips in the exact proportion needed by the paper mill. As a result, the paper mill can rapidly produce paper using the optimum recipe (i.e. lower production costs and more high-quality paper).

When wood chip sorting is implemented (Table 4), the supply chain profit is improved by up to 11%. The paper mill is the one that benefits the most from this change, sawmills' profit being the same or decreased slightly. The analysis based on collaboration benefit sharing will confirm the need for redistributing a part of the supply chain profit to ensure win−win relationship.

During the course of our study, one of the sawmills (i.e., sawmill C) started sorting its chip inventory in order to supply the paper mill more efficiently. This was not achieved using a specific equipment but rather via a different way to manage chip inventory all along the production process. Even though it is difficult to estimate the cost of this change, theoretical results confirm the impact it could have on the profit of the whole supply chain.

Scenario 4

With this scenario, we assume that companies take into account wood chip freshness. At each sawmill, production runs five days a week, generating both lumber and wood chips. A chip that is kept in stock from one to five days (i.e., one production week) is considered fresh regarding its moisture and its brilliance. When chips are in the system for six to ten days (i.e., two production weeks), we consider them less fresh. They are considered as not fresh when they have been kept in stock for at least 11 days. A fresher wood chip will involve a lower production cost and a greater paper quality. In particular, paper production cost can increase by up to 6.2% when chip quality is not adequate, encompassing

Table 5. Individual and supply chain profits obtained for Scenario 4.

	Scenario 4		
	RR	VMI	CPFR
Sawmill A	$3,414,390	$4,682,067	$4,670,937
Sawmill B	$2,727,762	$2,993,148	$2,947,063
Sawmill C	$6,797,901	$7,672,520	$7,669,053
Paper mill	$44,323,745	$45,321,093	$46,042,866
Supply chain	$57,263,798	$60,668,828	$61,329,919

longer set-ups and a higher use of chemical products. With these parameters factored into our model, the solution will always prioritize the use of fresh chips. For all the logistics strategies tested, the management of chip freshness contributes to increase supply chain profit by up to 31% (Table 5). It has a positive impact on the paper mill process by ensuring a better chip quality. On the other hand, it involves an effort for the sawmills that contributes to decrease their individual profit compared with the one obtained in previous scenarios when VMI or CPFR is implemented.

Managing freshness is a strategy these companies wanted to prioritize. However, theoretical results show that without a form of profit sharing, this strategy would only be beneficial for the paper mill.

Scenario 5

In Scenario 5, all five sawmills can supply wood chips to the paper mill. The volume delivered will, therefore, contribute to satisfy the paper mill demand and sawmill B will not have to be used to fill the supply gap (Table 6).

Results show that the supply chain profit improves significantly if all five sawmills supply chips to the paper mill. As mentioned, two sawmills were temporarily closed because of corporate decisions. Yet, our results indicate that if all five sawmills would operate within a coordinated system, decision-makers would find motivation to maintain operations at these units.

In summary, through our model, we have evaluated how profitable it would be for all sawmills and the paper mill to use VMI or CPFR rather than approaches like RR. Obviously, it involves a better management of inventories and a faster adjustment regarding demand variations. Above all, it requires agreement on a benefit sharing method.

Individual contribution and benefit sharing

Past research has demonstrated the complexity of implementing inter-firm collaborations, especially concerning the economies and benefits to share between members (see for example Audy et al. 2012). Some partners may bring to the partnership a new demand to satisfy, a higher volume to transport or a greater product quality, while others may have to change their process to contribute positively to the grand coalition. As a result, it becomes necessary to make sure that each partner will get a fair share of the collaboration benefits regarding its contribution to the partnership as well as its expectations.

This is why after identifying the profitability of each logistics strategy for both the sawmills and the paper mill, we tried to evaluate what share of the collaboration profit each

Table 6. Individual and supply chain profits obtained for Scenario 5.

	Scenario 5		
	RR	VMI	CPFR
Sawmill A	$3,427,123	$4,755,387	$4,793,497
Sawmill B	$2,727,007	$3,179,199	$3,172,092
Sawmill C	$6,798,938	$7,604,260	$7,581,468
Sawmill D	$45,180,397	$46,096,692	$46,423,414
Sawmill E	$1,154,400	$1,128,062	$1,167,282
Paper mill	$2,673,347	$2,491,850	$2,730,655
Supply chain	$61,961,212	$65,255,450	$65,868,409

Table 7. Share of the global profit obtained by each partner based on Scenario 4 with VMI.

	Optimization model	Shapley	Separable and non-separable costs	Kwasnica et al.
Sawmill A	7.72%	7.85%	7.06%	6.55%
Sawmill B	4.93%	18.56%	18.26%	6.55%
Sawmill C	12.65%	12.13%	11.42%	14.63%
Paper mill	74.70%	61.47%	63.26%	72.27%
Supply chain	100.00%	100.00%	100.00%	100.00%

player should get in regard to its contribution. Scenario 4 and the VMI approach for the three operating sawmills were used as a basis for the analysis. We chose this specific context because in our opinion, it seems a promising and likely strategy to increase the supply chain profit without necessitating a complete process reengineering. As explained previously, three methods were used to estimate the profit allocation, namely the Shapley value, separable and non-separable costs, and a method adapted from Kwasnica et al. (2005). Results appear in Table 7.

While the method adapted from Kwasnica et al. (2005) gives similar results to those obtained with the optimization model, results obtained with the Shapley value technique as well as the separable and non-separable costs are quite different. If we look at sawmill B, we can see that the profit computed from the optimization model is significantly lower than the part it should obtain regarding the Shapley value or separable and non-separable costs. Furthermore, results show that the paper mill gets a higher share of the profit than what it should obtain based on game theory methods. Because the wood chips produced at sawmill B is generally of higher quality than those from other sawmills, and since this sawmill is sometimes used to convert entire trees into chips to satisfy the paper mill demand rather than its lumber demand, its contribution to the overall profit is significant. Moreover, even though sawmill B is owned by the same company as the paper mill, these business units are typically managed independently and each plant will receive annual bonus depending on its performance. Sawmill B being mostly used to serve the needs of the paper mill, its annual bonus would certainly suffer from its weak presence on the lumber market. As a result, the paper mill should redistribute a part of its profit to this sawmill in order to take into consideration the effort made while ensuring a relationship beneficial for both parties. We have applied the same analysis for Scenario 4 with CPFR and obtained similar results.

Logistics strategy implementation and benefit sharing

When comparing the different logistics approaches tested in our research, we observed that VMI was the most profitable approach for sawmills while CPFR was the most beneficial strategy for the paper mill as well as for the supply chain. VMI involves a higher stock level at the paper mill to ensure a certain service level while with CPFR, sawmills have to maintain more stock at their production sites to synchronize supply chain operations

Table 8. Individual contribution of each partner based on Scenario 4 if implementation of CPFR as a replacement for VMI.

	Optimization model	Shapley	Separable and non-separable costs	Kwasnica et al.
Sawmill A	7.62%	7.71%	7.67%	7.63%
Sawmill B	4.81%	4.90%	4.83%	4.88%
Sawmill C	12.50%	12.62%	12.57%	12.51%
Paper mill	75.07%	74.78%	74.93%	74.98%

adequately and better use transportation capacity. Shapley value, separable/non-separable costs, and Kwasnica et al.'s method were again used to evaluate how the supply chain profit as well as the benefits generated from implementing CPFR could be shared amongst partners if CPFR was implemented as a replacement for VMI (Table 8).

Replacing VMI by a CPFR approach would require distributing a higher share of the profit to the five sawmill as compared to what they would obtained via an optimization model. CPFR benefits sharing would be the incentive to support the implementation of coordinated production planning and information sharing. Otherwise, sawmills will probably prefer to work together based on VMI, a simpler approach to implement.

Implementation challenges for the case study

It is interesting to link the theoretical analysis presented here to the reality of the case study. When we started to work with the companies, no formal collaboration scheme was being used. The region's economic situation was difficult and the companies started realizing that working together was required if they were to survive. Therefore, it was established that a weekly meeting involving directors of the sawmills, the director of the paper mill and a facilitator from the government would take place to discuss production plans, and timber volume availability. Decisions were still made individually, following a RR strategy. Using information discussed during these meetings, our experimentation yielded results that were presented to the facilitator. While impressed by the theoretical values obtained, the facilitator remained sceptical about the possibility to put VMI or CPFR into practice due to the level of inter-firm collaboration each would involve. At the same time, the companies stopped investing in the process and the facilitator was taken out. It has not yet been possible to test the implementation of our results but little doubts remain that collaboration would provide help to the industry in the Cote-Nord region. Implementing an approach like VMI would involve information sharing on a regular basis between the paper mill and the sawmills concerning wood chip inventory level, as they did before during their weekly meetings. This information sharing could be supported by a web platform or even emails at the beginning of the implementation. The price paid by the paper mill for wood chips could be used as the mechanism to redistribute a portion of the collaboration profit between partners (i.e., price for wood chips a little higher than the market price to reflect the portion of the profit to share). The price could next be adapted each year depending on the demand. Wood chips could be sorted as done by one of the sawmills, without necessitating any equipment acquisition. Measuring the freshness of the chips could be achieved via more efficient inventory management.

Concluding remarks

In this project, we have studied the case of five sawmills and one paper mill located in the Côte-Nord region in Quebec, Canada. Based on inter-firm relationship concepts, we have analysed how sawmills and the paper mill could better coordinate their operations to respond more efficiently to the paper mill demand and improve the overall supply chain profit. Three logistics approaches recognized in past research as efficient supply chain coordination mechanisms were selected for evaluation: RR, VMI and CPFR. The impact of each strategy on both individual and supply chain profits was tested for five different contexts. We have finally estimated the part of the profit each partner should get regarding its contribution to the collaboration, based on the Shapley value, separable and non-separable costs as well as a method adapted from Kwasnica et al. (2005). These methods were also used to estimate how the global profit should be shared if the most profitable approach for the supply chain (i.e., CPFR) was implemented as a replacement for VMI.

Results showed that a strong collaboration between the sawmills and the paper mill could contribute to increase the supply chain profit significantly. In particular, if sawmills are responsible for managing chip inventories at the paper mill site, their profit could increase by up to 44%. The gain is still significant if they decide to jointly plan supply chain operations using common forecasts. When looking at better ways of using the wood chip such as sorting species or managing its freshness, we observed a significant impact, especially for the paper mill. When the right wood chip is delivered to the paper mill with the right quality, the paper production process necessitates fewer adjustments so the production cost becomes lower. None of the approach was the most profitable one for both sawmills and the paper mill. While VMI was the strategy the most beneficial for sawmills, CPFR was the one that generated the highest profit for the paper mill. The analysis pointed out that the paper mill got the highest part of the profit with techniques like VMI or CPFR, but when considering the contribution of each partner, results showed the necessity to think about a way for redistributing the profit fairly between them.

With this research, we may observe that greater collaboration, even between competitors, can be profitable for the forest products companies under study. In the future, we would like to integrate collaboration both at the sawmill sites and in harvesting scheduling to deliver the right raw material at the best plant for maximizing supply chain benefits. We would also explore if the creation of an energy unit in the region could facilitate the management of the wood freshness. In this way, we could propose new ways of doing business that could help the forest sector to be more efficient.

Disclosure statement

No potential conflict of interest was reported by the authors.

ORCID

Luc LeBel (iD) http://orcid.org/0000-0001-9553-6139

References

Audy J-F, D'Amours S, Rousseau L-M. 2007. Collaborative planning in a log truck pickup and delivery problem. 6th Triennial Symposium on Transportation Analysis; June 10–15; Phuket Island.

Audy J-F, Lehoux N, D'Amours S, Rönnqvist M. 2012. A framework for an efficient implementation of logistics collaborations. Int T Oper Res. 19:633–657.

Beaudoin D, Frayret JM, LeBel L. 2010. Negotiation-based distributed wood procurement planning within a multi firm environment. For Policy Econ. 12:79–93.

Boyer M, Moreaux M, Truchon M. 2006. Partage des coûts et tarification des infrastructures [Cost sharing and infrastructure pricing], Monographie CIRANO 2006 MO-01 (2006) mars. Montréal: CIRANO.

Büyüközkan G, Vardaloğlu Z. 2012. Analyzing of CPFR success factors using fuzzy cognitive maps in retail industry. Expert Syst Appl. 39:10438–10455.

Cachon GP. 2003. Supply chain coordination with contracts. In: de KokAG, GravesSC, editors. Handbooks in operations research and management science. Amsterdam: Elsevier; p. 229–339.

Camarinha-Matos LM, Afsarmanesh H, Galeano N, Molina A. 2009. Collaborative networked organizations: concepts and practice in manufacturing enterprises. Comput Ind Eng. 57:46–60.

Cao M, Vonderembse MA, Zhang Q, Ragu-Nathan TS. 2010. Supply chain collaboration: conceptualisation and instrument development. Int J Prod Res. 48:6613–6635.

Chambost V, McNutt J, Stuart PR. 2009. Partnerships for successful enterprise transformation of forest industry companies implementing the forest biorefinery. Pulp & Paper Canada. 110:19–26.

Fabbe-Costes N, Lancini A. 2009. Gestion inter-organisationnelle des connaissances et gestion des chaînes logistiques: enjeux, limites et défis [Interfirm knowledge and supply chain management: issues and challenges]. Revue Management & Avenir. 29:123–145.

Frisk M, Göthe-Lundgren M, Jörnsten K, Rönnqvist M. 2010. Cost allocation in collaborative forest transportation. Eur J Oper Res. 205:448–458.

Iftekhar MS, Hailu A, Lindner RK. (2009). Comparisons of linear item pricing for iterative multiunit reverse combinatorial auctions. International Conference on Policy Modeling – EcoMod2009; June 24–26; Ottawa, Canada.

Jerbi W, Gaudreault J, D'Amours S, Nourelfath M, Lemieux S, Marier P, Bouchard M. 2012. Optimization/simulation-based framework for the evaluation of supply chain management policies in the forest product industry. 2012 IEEE International Conference on Systems, Man, and Cybernetics; October 14–17; Seoul, Korea.

Kwasnica AM, Ledyard JO, Porter D, DeMartini C. 2005. A new and improved design for multiobject iterative auctions. Management Science. 51:419–434.

Lefaix-Durand A, Robichaud F, Beauregard R, Frayret JM, Poulin D. 2006. Procurement strategies in the homebuilding industry: an exploratory study on the largest builders in the United States. J Forest Prod Bus Res. 3:1–22.

Lehoux N, D'Amours S, Frein Y, Langevin A, Penz B. 2011. Collaboration for a two-echelon supply chain in the pulp and paper industry: the use of incentives to increase profit. J Oper Res Soc. 62:581–592.

Lehoux N, D'Amours S, Langevin A. 2008. A win-win collaboration approach for a two-echelon supply chain: a case study in the pulp and paper industry. Eur J Ind Eng. 4:493–514.

Lehoux N, D'Amours S, Langevin A. 2014. Inter-firm collaborations and supply chain coordination: review of key elements and case study. Prod Plan Control. 25:858–872.

MacCarthy BL, Jayarathne PGSA. 2012. Sustainable collaborative supply networks in the international clothing industry: a comparative analysis of two retailers. Prod Plan Control. 23:252–268.

Marquès G, Thierry C, Lamothe J, Gourc D. 2010. A review of vendor managed inventory (VMI): from concept to processes. Production Planning & Control. 21:547–561.

Mason R, Lalwani C, Boughton R. 2007. Combining vertical and horizontal collaboration. Suppl Chain Manage An Int J. 12:187–199.

Myerson RB. 1997. Game theory, analysis of conflict. Cambridge: Harvard University Press.

Natural Resources Canada. 2015. Statistical data, forest resources. [Accessed September 21st,]. Available from: http://cfs.nrcan.gc.ca/statsprofile

Naud MP, D'amours S. 2011. Modèle de conception de chaînes logistiques vertes et collaboratives pour l'industrie des produits forestiers [Green supply chain design model for the forest products industry]. 9e Congrès International de Génie Industriel; Québec, Canada.

Okada N, Tanimoto K, Sakakibara H. 2009. Cost allocation. In: Hipel KW, editor. Conflict resolution, Vol. II. Encyclopedia of Life Support Systems. Oxford: Eolss Publishers.

Palander T, Vaatainen J. 2005. Impacts of interenterprise collaboration and backhauling on wood procurement in Finland. Scand J Forest Res. 20:177–183.

Ramanathan U, Gunasekaran A. 2014. Supply chain collaboration: impact of success in long-term partnerships. Int J Prod Economics. 147:252–259.

Simatupang TM, Sridharan R. 2008. Design for supply chain collaboration. Bus Process Manag J. 14:401–418.

Stadtler H. 2009. A framework for collaborative planning and state-of-the-art. Or Spektrum. 31:5–30.

Advances in profit-driven order promising for make-to-stock environments — a case study with a Canadian softwood lumber manufacturer

Rodrigo Cambiaghi Azevedo, Sophie D'Amours and Mikael Rönnqvist

ABSTRACT

Profit-driven order promising mechanisms have been receiving increasing attention from academia and practitioners in recent years. In spite of the recent advances in the field, its application in practice still remains a challenge. One of the most interesting approaches currently discussed is called aATP (allocated available-to-promise). In this paper, we propose some advances for current state-of-the-art aATP constructions in make-to-stock environments, mainly adapting it to commodity products sold through spot markets. Practical situations are tested through simulations supported by real data obtained from a major softwood manufacturer in Canada. Examples show that the proposed approach can increase the profitability of the producer.

1 Introduction

Activities associated with customer order promising (OP) are among the most important for manufacturing organizations (Kalyan 2002; Pibernik 2005; Chen et al. 2008; Meyr 2009; Alemany et al. 2013; Ali et al. 2014; Eppler 2015; Kilger and Meyr 2015). They directly impact the company's profitability and customer service level satisfaction in the short term and consequently, the medium and long-term competitiveness of the business (Kalyan 2002). Usually, service level satisfaction may be impacted by the response time to a customer inquiry as well as by the reliability of the promises in terms of order quantity and delivery date. OP activities also directly impact company profitability either by variables involved in each different order such as price, delivery mode alternatives, required product(s), payment flow, etc. or by decisions derived from the attempt to accommodate short-term in-balance between demand and available products or capacities. Typical OP decisions include: what orders to fulfil when supply is lacking, how to negotiate customer requirements (e.g. price, delivery date, product quantity), or even how not to accept serving certain customers. These decisions can become extremely complex in business contexts where, for instance, companies offer a large portfolio of products; they have a large number of customers; and there is an intense level of products phase-in and phase-out (Kalyan 2002).

To cope with this challenge in make-to-stock (MTS) environments, companies conventionally applied the concept of real-time available-to-promise (ATP) (Chen et al. 2008). When a customer requests an item, ATP allows companies to reply to the request consistently by quickly scanning all inventory locations to determine the stock position in time, which includes both on-hand inventories and non-allocated inventories to be produced. Usually, ATP functionalities also support associations between specific customers and certain inventory locations in the network as well as product substitution rules.

Although widely applied, traditional ATP constructions still fail to simultaneously optimize company profitability and customer service level strategy. When successfully applied, ATP promotes greater customer service level satisfaction by reserving inventory for accepted orders. However, it does it in a First-In-First-Out (FIFO) manner, ignoring the opportunity to optimize the company's profitability through heterogeneous customer willingness-to-pay and alternatives of value chain costs.

To address this gap, the concept of aATP (allocated available-to-promise) has emerged. The aATP procedure is a two-step (Figure 1) where the first, 'ATP Allocation' or 'Allocation Planning', aims to assign scarce resources, i.e. not yet reserved stock and planned production quantities, to different customer segments, which are generally clustered based on their profitability or other priority measures (Quante et al. 2009b; Meyr 2009). The main issue of this procedure lies in how to allocate capacities to each customer segment ahead of orders arrivals. To do so, two mechanisms may be applied: rule-based ATP allocation and optimization-based allocation models. Kilger and Meyr (2015) discuss the first by presenting simple rules that are usually applied in Advanced Planning System (APS) systems. These rules quote an overall ATP quantity to different customer classes on the basis of priority rankings, with respect to some pre-defined fixed shares or proportional to the original forecasts of different customers or markets (Meyr 2009). Rule-based ATP

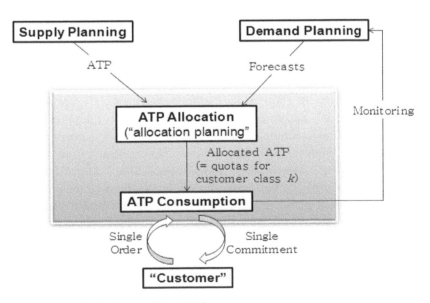

Figure 1. aATP two-step logic (source: Meyr 2009).

allocation mechanisms are simple and straightforward, which leads to relative simplicity in implementation and integration with other business functions (Ball et al. 2004). Optimization-based allocation models are more complex, but they allow better representation of business environments. They can explicitly take into account variations in profitability at the customer order level, complex resource constraints, interactions between similar products, etc. (Ball et al. 2004).

The second step of the aATP logic consists in the 'ATP Consumption.' Once segment quotas are established and an order is received, the OP system checks the allocations for the corresponding order class; if ATP is available, ATP can be consumed and the order is quoted accordingly. Otherwise, the system searches for further options to fulfil the demand (*e.g.* later delivery date or alternative product) or reject the order.

In addition, once orders start to arrive, the consumption of each customer segment capacity allocation must be measured through continuous monitoring. Differences in consumption in association with the forecast can be used to review the allocations (step 1) (Kilger and Meyr 2015). It is in the ATP consumption step that advanced mechanisms are used to manage product/service availability (Eppler 2015). One of the most common types of control is the concept of booking limits (BLs). BLs, commonly applied in the service industry, are controls that limit the amount of capacity that can be sold to any particular class at a given point in time. BLs can be either partitioned or nested. A partitioned BL divides the available capacity among separated classes, not allowing the classes to share the same capacity available. With nested BLs, the capacity available to different classes overlaps in a hierarchical manner, with higher-raked classes having access to all of the capacity reserved for lower-ranked classes (Talluri and Ryzin 2005). This paper explores the application of nested BLs for customer orders where request delivery dates are in the future.

aATP constructions should not be viewed solely as an optimization model contribution but rather as a decision support system used to maximize profit over the planning horizon. This is accomplished by deciding in an environment with high numbers of orders, different customer behaviours, and multiple options of value chain costs, whether to accept or to reject a given order in anticipation of more profitable future orders.

1.1 Literature review on aATP applications

The concept of aATP has its roots in the yield management (or revenue management) body of knowledge which has emerged based on its successful development in service industries such as airlines and hotels (Talluri and Ryzin 2005; Quante et al. 2009b; Meyr 2009; Quante et al. 2009a). Although similar, the application of aATP in MTS manufacturing settings is differentiated from the classical airline problem by explicitly considering exogenous past and future replenishment as well as by the possible consideration of inventory holding costs and backlogging costs (Quante et al. 2009a). In the service industry, capacity is assumed perishable and fixed.

Even though, for more than two decades, there has been frequent questioning in the literature about how to effectively accept or deny orders in manufacturing settings (Guerrero and Kern 1988) and, during the same period of time, the development and application of traditional revenue management concepts have been intensively debated, only very recently have researchers been able to simultaneously explore these two main streams for MTS manufacturers (Vogel 2013; Eppler 2015). In contrast, the application of revenue management concepts in 'to-order' environments (i.e. make-to-order (MTO) and assemble-to-order (ATO)), has more often been found in the literature (*e.g.* Tamura and Fujita 1995; Kalyan 2002; Venkatadri et al. 2006; Chen et al. 2008; Yang and Fu 2008; Chiang and Wu 2011), a fact that, according to Quante et al. (2009a), might be related to the intuitive correspondence between production capacities and perishable assets. Likewise, Harris and Pinder (1995) affirm that many revenue management environment characteristics are also found in ATO environments such as perishable resources, fixed capacity, high-capacity changing costs, demand segmentation, advance sales/bookings, stochastic demand, and historical sales data and forecasting capability.

In 2009, Meyr (2009) presented a pioneer work for aATP construction in MTS settings. Using deterministic linear programming (LP) models for ATP allocation and ATP consumption, the author applies the aATP concept to the lighting industry where bulbs, fluorescent lamps etc. are produced to stock based on medium-term demand forecasts. Considering one single product, Meyr (2009) is able to demonstrate the benefits of the concept compared to First-In-First-Served (FIFS) and batch order processing approaches by testing two different levels of aggregation for segmented allocated volumes. In a detailed version, named $aATP_{kt\tau}$, the allocation of ATP specifies not only the period t that the volume becomes available for the priority class k, but also indicates in which period τ it should be consumed. In the aggregated version, named $aATP_{kt}$, ATP values are relaxed in terms of the period τ when the allocations can be consumed. In addition, simulations analyse the application of these different levels of aggregation with diverse rules to access allocations among classes (access just to the original class, to all lower priority classes, to higher priority classes only or to all existing classes) as well as two different search mechanisms (free among the different classes or with a pre-defined searching sequence). The analyses emphasize the advantages of the temporal reservation obtained through detailed allocation mechanisms with access to lower priority classes and the use of pre-determined searching sequences. Finally, the author simulates the impact of the application considering different numbers of priority classes and concludes that profit can be increased by introducing more priority classes as long as different customer orders show various per unit profits.

Using stochastic modelling for demand and due date information and deterministic approach for inventory and replenishment data, Pibernik and Yadav (2009) propose a two-step aATP model that differentiates between two customer classes for one single product manufactured in an MTS environment. In an innovative approach, the authors utilize a target service level (instead of expected profits) as the basis for determining inventory allocations.

Similarly, Quante et al. (2009a) develop an aATP approach where demand for different customer classes presents stochastic behaviour while inventory replenishments are exogenous and deterministically known. The main difference between the two propositions lies in the fact that Quante et al. (2009a) optimize their developments based on expected profit, instead of service level target as applied by Pibernik and Yadav (2009). In addition, the model presented by Quante et al. (2009a) also allows backlogging demand in case of shortages; a fact that increments the applicability of the model. Through massive simulation efforts against FIFS and deterministic aATP formulations, the authors are able to draw interesting analyses regarding the impact of demand variability, customer heterogeneity and supply shortage in the outputs of the mechanism. However, it is worth mentioning that the model's main disadvantage might be its limited scalability due to computational demand for large problems (Quante et al. 2009a).

Most recently, Eppler (2015) developed stochastic linear programming (SLP) models for both allocation planning (AP) and order fulfilment decisions. Eppler (2015) also performed interesting numerical studies for single-period and multi-period AP processes. However, the study focuses on a single-product and single-inventory source environment and does not explore existing dynamics in the commodity business (e.g., price fluctuation along the planning horizon). Similarly, Vogel (2013) discusses different profit-based allocation schemes for heterogeneous, multi-stage customer hierarchies in MTS. In spite of focusing on non-commodity settings, the study provides interesting insights into customer segmentation and demand forecasting challenges.

1.2 Contribution and organization of the paper

The application of the aATP concept in MTS manufacturing environments has been capturing increasing interest due to its potential benefits demonstrated in diverse constructions. Following this trend, this paper aims to contribute to the progression of the aATP concept for MTS environments by analysing its application in a commodity business where products are sold through spot markets. In order to do so, it extends the model and analyses presented by Meyr (2009) in four central ways: first, due to the nature of the softwood lumber business, a network perspective has been added on the construction of the model. The proposed model takes into consideration several storage locations and different geographical customer segments. By applying this perspective, nested BLs for different customer segments can be developed based on the combination of different price and cost behaviours along the network. In addition, the network perspective allows the application of an optimization engine capable of determining the most profitable sourcing location every time a new order is promised (allocation consumption step). The use of the engine replaces the static assignment client — storage location found in traditional APS systems.

Second, the study presents the first evidence in the literature of aATP performance when customer requests more than one product in the same order. Intuitively, an environment where orders carry multiple products may create a lesser degree of freedom for the allocation consumption step since it must combine products within orders that were allocated based on independent forecasts. This analysis might show evidence of a weakening in the performance of aATP solutions.

Third, by reflecting a commodity-product in spot market, the model, differently from the proposition presented by Meyr (2009), needs to be able to explicitly treat the dynamicity of market price as well as behaviour for customers' willingness to pay. Due to the transactional nature of a commodity spot market, in contrast to contract-based relationships, customers cannot be statically classified into categories of profitability or willingness to pay. The order-promising logic needs to allow a client to present different profiles of willingness to pay in different negotiations depending on its specific situation (e.g. low level of inventory and visibility of high demand in the short-term) and the market's condition. By accepting the dynamicity of the environment, demand should be forecasted by different behaviours of customer willingness to pay and product.

Finally, Meyr (2009) affirms that a temporal reservation of inventory would generally be advantageous, but its practical application is only reasonable if customer demand can be forecast reliably enough and he calls for further research where lower forecast accuracy should be tested. Using exemplary tests, the paper debates the feasibility of using aATP in environments, such as in the spot market for softwood lumber, where demand and price forecast present considerable degree of uncertainty even with horizons as short as from one to four weeks.

The remaining of this paper is organized as follows: first, the AP and the real-time OP models are presented and debated. Section 3 analyses the application of the proposed models in a case study with a softwood lumber manufacturer. Finally, the main findings and opportunities to advance future research are outlined in Section 4.

2 Real-time order promising using aATP, booking limits and sourcing location optimization in distributed networks

In an MTS commodity-type business, a company usually produces L products in M different manufacturing locations and sells them in S customer/market segments. Each manufacturing site determines its production schedule based on medium-term forecasts and short-term resources optimization. For this reason, aATP models applying exogenous replenishment are considered appropriate. In addition, different customer classes present different price and demand behaviours which may vary along the planning horizon. Transportation costs between the inventory locations and the target markets as well as cross-border taxes and exchange rates fluctuations directly impact the profitability of each segment in time, and consequently, decisions on how much and when to sell for each segment. In this environment, a key challenge is to try to maximize profit over a period of time by deciding whether to accept or reject an order from a given customer.

To address this problem, the aATP logic is proposed. Its construction is defined by two main steps. First, an optimization model allocates the available inventory (on-hand and planned) to the different customer segments in order to optimize the company's total expected net profit. The model is based on deterministic information concerning forecasts (prices and demand), value chain costs for different customer segments as well as general planning data (i.e. export quotas). Allocations are thus nested within BLs following the logic that, in case demand exceeds the original forecast, more profitable customers have priority over allocations for less profitable customers. This mechanism aims to help protect the firm's profitability against demand forecast errors. It is important to emphasize that although the paper analyses the impact of different forecast error profiles in the overall performance of the model, the use of deterministic data for demand and price is still one of the main limitations of the proposed construction. This limitation will be addressed in Section 4 as a key recommendation for future research.

In a second step, as customer orders arrive, BLs availabilities are checked, and if available, consumed according to an optimization algorithm that searches for the most profitable sourcing location. Figure 2 depicts the logic proposed. While for the AP step a LP problem is proposed, for the OP problem a mixed integer programming (MIP) model is applied. The description of each algorithm is presented afterward.

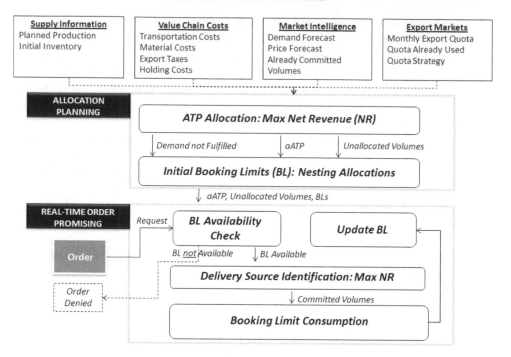

Figure 2. Proposed order promising scheme.

2.1 Step 1: Allocation planning (aATP and initial booking limits calculation)

As mentioned earlier, the AP step aims to assign the available inventory and planned production volumes to the different customer segments in order to optimize the company's total expected net profit. Table 1 describes all sets, parameters, coefficients and decision variables involved in the model. Next, the objective function (1) and the model constraints regarding the supply flow balance (2), the demand flow balance (3) and the export monthly quota (4) are given as follows:

$$\text{Maximize} \sum_{l=1}^{L} \sum_{m=1}^{M} \sum_{t=1}^{T} \sum_{s=1}^{S} \sum_{v=1}^{V} n_{lmtsv} * x_{lmtsv}. \tag{1}$$

Subject to

$$\left(\sum_{s=1}^{S} \sum_{v=1}^{V} x_{lmtsv} + w_{lmtsv} \right) + y_{lmt} = r_{lmt}, \quad t=1,\ldots,T; \quad m=1,\ldots,M; \quad l=1,\ldots,L \tag{2}$$

$$\sum_{m=1}^{M} \sum_{t=1}^{T} (x_{lmtsv} + w_{lmtsv}) + s_{lsv}, = d_{lsv} \quad l=1,\ldots,L; \quad s=1,\ldots,S; \quad v=1,\ldots,V \tag{3}$$

$$\sum_{l=1}^{L} \sum_{m=1}^{M} \sum_{s \in S_e} \sum_{t=1,}^{T} \sum_{v \in V_k} (x_{lmtsv} + w_{lmtsv}) \leq g_k, \quad k=1,\ldots,K \tag{4}$$

$$x_{lmtsv} \geq 0, \quad l=1,\ldots,L; \quad m=1,\ldots,M; \quad t=1,\ldots,T; \quad s=1,\ldots,S; v=1,\ldots,V \tag{5a}$$

$$y_{lmt} \geq 0, \quad l=1,\ldots,L; \quad m=1,\ldots,M; \quad t=1,\ldots,T \tag{5b}$$

$$s_{lsv} \geq 0, \quad l=1,\ldots,L; \quad s=1,\ldots,S; \quad v=1,\ldots,V. \tag{5c}$$

Table 1. Sets, parameters, coefficients and decision variables used by the AP model (Step 1).

Sets/indices	
$m = 1, ..., M$	Manufacturing sites
$s = 1, ..., S$	Customer segments
$l = 1, ..., L$	Products (SKU)
$t = 1, ..., T$	Time buckets — production (week)
$v = 1, ..., V$	Time buckets — consumption (week)
$k = 1, ..., K$	Months involved in the allocation horizon
S_e	All export segments (customers)
V_k	All planning weeks belonging to month k
Parameters	
a_{ms}	Unit transportation cost between plant m and segment s
b_s	Export tax for segment s (% of sales price)
c_{lmt}	Material cost of product l produced at plant m in week t
d_{lsv}	Forecasted demand of product l for segment s in week v
f_{ltv}	Unit holding costs to allocate a product l produced at week t in week v
p_{lsv}	Price of product l for segment s in week v
i_{lm}	Initial inventory of product l at plant m
r_{lmt}	Planned production of product l at plant m in the week t
q_k	Quota for export market in the month k
e_k	% of the quota for month k considered in the planning cycle
u_k	Volume of export quota for month k already consumed by previous planning cycles
w_{lmtsv}	Allocations (of product l at plant m in the week t to be used by segment s in consumption period v) already committed in previous planning cycles
Coefficients	
n_{lmtsv}	Net Profit for product l from plant m produced at week t, allocated to segment s to be consumed at week v $((p_{lsv}*(1 - b_s)) - f_{ltv} - a_{ms} - c_{lmt})$
g_k	Available export quota for month k and set S_e $e_k*(q_k - u_k)$
Decision variables	
x_{lmtsv}	Volume of product l from plant m produced in week t allocated to segment s for the week v ($t \leq v$)
y_{lmt}	Unallocated production of product l produced at plant m on week t
s_{lsv}	Not fulfilled expected demand of a product l from segment s within week v

The objective function (1) allocates available products to customer segments in order to optimize the company's total expected net profit over the planning horizon. To do so, the model considers four set of constraints: in constraint set (2), the supply flow is balanced once the total volume allocated (x), plus the volumes previously committed (w) and the unallocated volumes (y) must be equal to the volume current in stock (i) (t = 1) plus the planned production (r). Constraint set (3) balances the demand flow by guaranteeing that the summation of the volume allocated (x) and the amounts previously committed (w) in all factories and production weeks plus the volume of demand not fulfilled (s) must be equal to the total demand for the planning horizon (d). Constraint set (4) is applied in order to guarantee that the volumes allocated to export segments (S_e) must be lower or equal than the total export quota for each month involved in the planning horizon (V_k). Finally, constraint set (5) assures that all variables are non-negative. The problem is formulated in LP model.

Once the volumes x are allocated to each segment, a special feature is used in order to apply the concept of BLs. The BL's are set by product, plant, segment and consumption week. The BL of a product l for the segment s in manufacturing m in week v is calculated by nesting four sorts of allocation:

$$BL_{lsmv} = B1_{lsmv} + B2_{lsmv} + B3_{lsmv} + B4_{lsmv}. \tag{6}$$

(1) *Exact allocations (6a):* all volume for product l available in manufacturing m allocated specifically to segment s in the week v:

$$B1_{lsmv} = \sum_{t=1}^{T} (x_{lmtsv}). \tag{6a}$$

(2) *Unallocated volumes (6b):* volumes eventually not allocated to any segment must be considered within all BLs. In other words, non-allocated volumes should be assigned to fulfil any order when original allocation for the segment is not sufficient. Sub-indices s and v are fixed according to the incoming non-forecasted customer order:

$$B2_{lsmv} = \sum_{t=1}^{T} y_{lmt}. \tag{6b}$$

(3) *Different consumption weeks reallocation (6c):* at this stage the model searches for volumes allocated to weeks other than v and checks their expected net profit against the one which would be achieved by reallocating the volume for week v. For weeks when their expected net profit (n) is lower than the one at week v, their volumes are thus considered available within the BL of week v. Therefore, if an order for segment s and consumption period v cannot be fulfilled with a predefined allocated volume, then the model tries to confirm the order by searching for volumes allocated to less profitable weeks that will be available on or prior to consumption week v. This mechanism avoids volumes being held for less profitable weeks when deliveries are requested by more profitable ones:

$$B3_{lsmv} = \sum_{t=1}^{T}\sum_{\bar{t}=1}^{T} (x_{lmts\bar{t}})\bar{t} \neq v; \ t \leq \bar{t}; \ n_{lmtsv} \geq n_{lmts\bar{t}}. \tag{6c}$$

(4) *Different segments reallocation (6d):* in this subsection, volumes allocated for less profitable segments, considering all possible weeks, are searched and their volumes inserted into the more profitable BL. Consequently, if an order for segment s and consumption week v cannot be fulfilled with the allocation defined in Step 1 (AP), the model tries to confirm the order by searching for volumes in less profitable segments that will be available on or prior to consumption week v. This procedure avoids orders from more profitable customers being denied while volumes are held for less profitable ones:

$$B4_{lsmv} = \sum_{t=1}^{T}\sum_{\bar{s}=1}^{S}\sum_{\bar{t}=1}^{T} (x_{lmt\bar{s}\bar{t}})\bar{s} \neq s; \ t \leq \bar{t}; n_{lmtsv} \geq n_{lmt\bar{s}\bar{t}}. \tag{6d}$$

2.2 Step 2: Real-time order promising model

Once the AP step has been accomplished, customer requests start to arrive in a random fashion; one at a time. At this moment, a real-time OP mechanism is required in order to

Table 2. Additional parameters, sets, and decision variables used by the ATP consumption model (Step 2).

Additional sets	
$B_{oi}^{\bar{v}}$	Set of different consumption weeks ($\bar{v}$) that are included into the booking limit once they generate lower or equal net profit $\bar{v} \neq i; t \leq i;\ n_{lmtoi} \geq n_{lmto\bar{v}}$
$B_{oi\bar{v}}^{\bar{s}}$	Set of different customer segments ($\bar{s}$) that are included into the booking limit once they generate lower or equal net profit. $\bar{s} \neq o; t \leq i;\ n_{lmtoi} \geq n_{lmt\bar{s}\bar{v}}$
Additional parameters	
$\bar{x}_{lmtsv}$	Current available volume of product l from manufacturing site m produced at week t allocated to segment s for week v
$\bar{y}_{lmt}$	Current available non-allocated volume of product l produced at manufacturing site m in week t
γ_1	Parameter to set the booking limit consumption priority for non-allocated volumes
γ_2	Parameter to set the booking limit consumption priority for allocated volumes of the same segment but for different consumption weeks
γ_3	Parameter to set the booking limit consumption priority for allocated volumes from other segments
h_l	The quantity required for product l
i	The consumption week requested by the order
o	The segment requesting the order
j	The number of different products requested by the order
u_v	Cumulative volume already accepted for export markets for week v.
Decision variables	
x_{lmtsv}^a	Variable that accounts for from which allocation x the order will be fulfilled ($x_{lmtsv}^a = 0,\ \forall t > i$)
y_{lmt}^a	Variable that accounts for from which allocation y the order will be fulfilled ($y_{lmt}^a = 0,\ \forall t > i$)
z_m	Binary variable that defines from which manufacturing site a product(s) will be delivered

cope with the objective of net profit maximization. Table 2 summarizes all new required sets and parameters, the solution from the previous step as well as decision variables. In this step, all products of an order must be withdrawn from the same manufacturing site and hence binary variables need to be applied. Subsequently, the problem is formulated in an MIP model as follows:

$$
Maximize \sum_{l=1}^{L}\sum_{m=1}^{M}\sum_{t=1}^{T} n_{lmtoi} * x_{lmtoi}^a + \gamma_1 * \left(\sum_{l=1}^{L}\sum_{t=1}^{T}\sum_{m=1}^{M} n_{lmtoi} * y_{lmt}^a \right)
$$

$$
+ \gamma_2 * \left(\sum_{l=1}^{L}\sum_{m=1}^{M}\sum_{(t,\bar{v})\in B_{oi}^{\bar{v}}} n_{lmtoi} * x_{lmto\bar{v}}^a \right) + \gamma_3 * \left(\sum_{l=1}^{L}\sum_{m=1}^{M}\sum_{(\bar{s},t,\bar{v})\in B_{oi\bar{v}}^{\bar{s}}} n_{lmtoi} * x_{lmt\bar{s}\bar{v}}^a \right) \tag{7}
$$

Subject to

$$
x_{lmtsv}^a \leq \bar{x}_{lmtsv} \ \forall\ l,\ m, t, s, v \tag{8a}
$$

$$
x_{lmtsv}^a \geq 0 \ \forall\ l,\ m, t, s, v \tag{8b}
$$

$$
y_{lmt}^a \leq \bar{y}_{lmt} \ \forall\ l, m, t \tag{9a}
$$

$$
y_{lmt}^a \geq 0 \ \forall\ l,\ m, t \tag{9b}
$$

$$
\sum_{t=1}^{T}\sum_{s=1}^{S}\sum_{v=1}^{V} x_{lmtsv}^a + \sum_{t=1}^{T} y_{lmt}^a = h_l * z_m \ \forall\ l, m \tag{10}
$$

$$
\sum_{m=1}^{M} z_m \leq 1 \tag{11}
$$

$$
z_m\ binary,\ \forall\ m \tag{12}
$$

$$
\sum_{l=1}^{L}\sum_{m=1}^{M}\sum_{s\in S_e}\sum_{t=1}^{T}\sum_{v\in V_k} x_{lmtsv}^a + \sum_{l=1}^{L}\sum_{m=1}^{M}\sum_{t=1}^{T} y_{lmt}^a + \left(\sum_{v\in V_k} u_v \right) \leq g_k. \tag{13}
$$

As noticed, the BL logic is translated into the objective function described above in (7). This fact facilitates the simulations of the scenarios analysed. In this way, the objective function is composed of four different allocation consumption formulas which represent the logic of the BL construction (6a, 6b, 6c and 6d). The parameters γ_1, γ_2 and γ_3 set the BL consumption priority when demand exceeds the exact allocation.

Constraint sets (8) to (12) impose four set of business constraints: Initially, volumes to fulfil an order, x^a and y^a, must always be lower or equal to the current available allocation $\bar{x}$ (8a) or non-allocated volumes $\bar{y}$ (9a); moreover, they must always assume non-negative values (8b and 9b). The following set of constraints (10 and 11) impose that the volume required for all products within an order comes from the same sourcing location and that they all must be either accepted or denied, avoiding partial fulfilment. Constraint set (12) is the binary restriction. Constraint set (13) is applied in the case where the requesting customer belongs to an export segment ($s \in S_e$). In this circumstance, the volume required by the customer order (x^a and/or y^a) plus the cumulative volume already accepted for S_e for all weeks belonging to month k ($v \in k$), must be equal or lower than the available export quota for month k.

3 Case study with a Canadian softwood lumber manufacturer

The proposed model was tested through a case study from a softwood lumber manufacturer located in Eastern Canada. The forest product industry constitutes one of Canada's main manufacturing industries, representing approximately 5.3% of Canada's total employment (Lehoux et al. 2012; Ali et al. 2014). Although several research studies on supply chain management topics can be found in the industry (i.e., Donald et al. 2001; Forget et al. 2008; Marier et al. 2014), the application of revenue management concepts is still in its infancy.

Lumber manufacturers located in Eastern Canada traditionally produce in an MTS mode and distribute their products primarily to the eastern Canadian and northeastern US markets. They sell their products predominately through two types of relationships; contracts and in the spot market. Our investigation in this paper lies upon the latter. Macro-economic forces such as the fierce international trade competition encountered with the introduction of new exporter countries; the rise of energy costs in recent years; the unstable commercialization scenario created due to the intensive trade dispute with US government; and the severe downturn in the US housing market have recently created an ideal situation for the introduction of new business models for lumber manufacturers where revenue management concepts could play a significant role.

Similar to Meyr (2009), in order to evaluate the proposed aATP models, three additional formulations were required: Global Optimization (GO), aATP without BLs, and FIFS. GO is considered the ideal proposition once it optimizes the company results by allocating supply to demand ex-post for all orders arriving within the planning horizon. The construction 'aATP without booking limits' (aATP-FIFS) uses the aATP models developed without making use of the BL feature. It serves customers exclusively from their respective allocated volumes in an FIFS manner; in this way allowing analyses on the impact of using the BL feature. Finally, the FIFS strategy does not reserve volumes in anticipation of different demand classes. It simply confirms order requests based on the availability of the product from its most profitable sourcing location in the supply network.

One might question the current practices of OP in the softwood lumber industry. Taking into consideration the company used for the case study described here, as well as the experience of our research teams through several other engagements in this industry (i.e., Lehoux et al. 2012; Ali et al. 2014; Marier et al. 2014,), it can be asserted that forecast information such as price, demand, supply chain costs and exchange rates, are not properly structured to support profit optimization when promising orders to customers. Lumber manufacturers do have 'good feelings' about customer average price sensitiveness and current supply chain costs (Johansson 2004); however, they try to optimize profitability on an order-to-order basis using indexed market price of the past week as reference. In this way, no controlled protection is offered to higher margin customers and revenue displacements are rarely managed and controlled properly. In other words, the current industry practice focuses on the optimization of weekly average order prices against market reference prices and not on the global allocation of available resources to different customer segments forecasts (demand and price) within a longer time horizon. Therefore, the benefits of the proposed models are compared against the additional constructions (GO, aATP-FIFS and FIFS) and validated with members of the company under study.

All models were developed in the modelling language AMPL version 11.2 using the CPLEX solver. All experiments are executed on a Windows Platform using Intel Core Duo workstation with CPU 1.33 GHz, 2GB of RAM, and Windows Vista Business Edition Version 2007.

3.1 Experiment data

In this section, the main data utilized in the experiment are introduced.

First of all, it is important to emphasize that, in terms of consumer markets, the case company basically sells in two territories: the US Great Lakes zone (GL), where the reference location is Columbus (Ohio) and the eastern Canadian market (CAN), where the reference location is Montreal (Quebec). Reference locations are important in commodity businesses as they represent the geographical point where prices and product shipping responsibilities are agreed upon. Trade between lumber manufacturers and their customers considers 'Cost, Insurance and Freight' (CIF) shipping method to the geographical reference location and 'Free-on-Board' (FOB) to the order final destination.

Therefore, to enable the application of the aATP proposition, an initial effort analysed the company's customer base in terms of its willingness to pay. Using knowledge of the sales team, independent customer segments were organized. Originally, the segmentation respected the market geographical division (GL and CAN) mainly due to their different price behaviour, the fluctuation of the US and Canadian dollar exchange rate, and the dissimilar transportation costs involved in each market. However, within each market an analysis in terms of price sensitiveness for each customer was feasible. Customer price sensitivity was used as an indicator of willingness to pay (lower sensitivity = higher willingness to pay; higher sensitivity = lower willingness to pay). Further improvements in terms of customer segmentation will be reinforced later in this paper.

Using sales data and the experience of the sales team of the case company, the segmentation analysis showed that the majority of customers (62% in terms of number of

Table 3. Price segmentation analysis by number of products and total volume.

Price sensitiveness	Number of customers	Total volume
Lower than average market price	30%	19%
Average market price	62%	72%
Higher than average market price	8%	9%

Table 4. Market segments applied in the experiment.

Eastern Canada	Seg. 1 — Premium price	Seg. 2 — Market price	Seg. 3 — Below market price
US Great Lakes	Seg. 4 — Premium price	Seg. 5 — Market price	Seg. 6 — Below market price

customers and 72% of the volume sold) have average price sensitiveness when negotiating an order, which might be understood as that, on average, their willingness to pay might be centred around market price. In addition, 30% of the customers (19% in terms of volume) aims to disburse a lower than market price during the negotiation (higher price sensitivity). However, 9% of the volume (8% in terms of number of customers) might be sold at a premium price once customers present lower price sensitivity. Table 3 summarizes the numbers for the different segments of the experiment. This fact indicates that, in a situation where demand exceeds supply, premium price customers should be prioritized to the detriment of more highly sensitive ones in order to optimize the company's total revenue. Along with this analysis and the two distinct geographical markets, the customer base was divided into six segments as depicted above (Table 4).

The experiment setting was further structured considering some conditional aspects. First, three products were selected for the simulations. This number of products allows investigations considering multi-products interrelationships without setting a severe level of complexity for data collection, management and analyses. Second, it considered a single transportation capacity in full truckload (TL) strategy. Further research might employ different capacities and strategies in order to provide better evidence of the impact of transportation costs optimization. In addition, similar to Pibernik and Yadav (2009), orders received are always assumed to be requested in one full truck-load volume. In addition, the models also do not consider volume negotiation or product substitution schemes. Finally, although the planning horizon spans four weeks, optimization models were simulated considering orders received during the first week. Orders arriving during any week are assumed to present order delivery dates distributed along the next four weeks in the planning horizon according to a profile of delivery date request obtained from the manufacturer (Tables 5–7). This condition is possible due to the myopic customer behaviour adopted, which assumes that order decisions are instantaneous and definitive. In other words, once an order is denied, its demand does not return as a request for a future period. Thus, considering any period between two AP activities is sufficient to identify the impact in the company's profitability for the different OP mechanisms.

Table 5. Order entry behaviour — number of orders and TLs for each product in all delivery weeks.

Products	No. of orders and no. of TLs per delivery week				
	W1	W2	W3	W4	Total
1	5	8	8	5	26
2	5	8	5	8	26
3	8	9	6	6	29
Total	18	25	19	19	81

Table 6. Order entry behaviour — profile by customer segment and consuming weeks in terms of number of orders and volume (truckload).

Segments	No. of orders and no. of TLs per week				
	W1	W2	W3	W4	**Total**
1	3	3	3	2	11
2	6	7	7	6	26
3	3	3	3	2	11
4	1	3	1	2	7
5	4	6	4	5	19
6	1	3	1	2	7
Total	18	25	19	19	81

Table 7. Order entry behaviour — profile by customer segment and products in terms of number of orders and volume (truckload).

Segments	No. of orders and no. of TLs per product			
	L1	L2	L3	Total
1	3	4	4	11
2	7	8	11	26
3	3	4	4	11
4	3	2	2	7
5	7	6	6	19
6	3	2	2	7
Total	26	26	29	81

In terms of value network, at the time of the analyses, the company had two sawmills situated in different locations in Eastern Canada. Inventories are associated with each of these sawmills and no reload centres or warehouse locations were considered in the value chain represented.

The planning horizon adopted for the experiment considered information from May 19 and June 13. In the lumber business, manufacturers usually revise the expected market prices before a new week starts. Aligned to that, Figure 3 illustrates the experiment

Figure 3. Illustrative description of the order entry environment addressed in the experiment.

simulating the AP activity on May 16 and subsequently the receipt and decisions taken for the orders arriving between May 18 and 23.

All information required for the tests (e.g. sales and price forecasts, value chain costs, etc.) were reconstructed with and validated by the company's sales members. In terms of export costs, a charge of 15% over sales price was applied for all US segments according to the *Softwood Lumber Products Export Charge Act*, agreed on between the US and Canadian governments in November 2006. In terms of BL consumption strategy, the priority sequence first checks for non-allocated volumes and subsequently for the most profitable option independently if it comes from a different consumption week for the same customer class or from different segments reallocations. This research simply used $\gamma_1 = 0.01$ and $\gamma_2 = \gamma_3 = 0.001$. Further investigation is required to determine if the normalization of these values impacts the optimal performance of the model.

Finally, an order request database was constructed with the purpose of defining the order entry behaviour for the selected products during one week. Based on the order request profile presented in Tables 5–7, 81 orders were selected. These orders were used to develop seven scenarios according to different orders arrival sequence. The scenarios (different sequences of the same set of orders) were randomly generated and their developments intended to capture different performances from the proposed constructions.

3.2 Benefits of aATP compared to FIFS

Initially, considering one product per order and 100% forecast accuracy in the AP step, simulations indicate that GO, aATP-BL and aATP-FIFS present the same performance in terms of achieved net profit and total number of orders accepted. The reason for this outcome is that in a completely predictable environment, the ATP allocation step is able to select the most profitable orders when building the customer class quotas and, consequently, BLs are not even necessary. For this reason, results for GO and aATP-FIFS will be omitted from the tables in this section.

However, in this environment, simulations show that the final outcomes achieved by the aATP-BL (and also GO and aATP-FIFS) construction exceeds the performance of the FIFS model in all scenarios analysed. While for all scenarios the aATP-BL generates a steady outcome of $380,847, the FIFS varies from its best result (scenario 4) of $354,186 and in its lower performance (scenario 7) by $349,136. Table 8 depicts the results of both models for the seven scenarios analysed.

The reason for the higher performance achieved by the aATP-BL propositions lies in the double effect of accepting a higher number of orders as well as selecting orders from more profitable segments. Table 9 shows the number of orders denied by each construction in each scenario analysed. While using the aATP-BL model a company would deny, independently of the orders arrival sequence, 25 orders (out of 81), by applying FIFS it would deny either 27 or 28 orders (DIFFERENCE = 2 or 3, respectively).

Table 8. Experiment net profit results for aATP-BL and FIFS.

Scenarios	1	2	3	4	5	6	7	Std dev.
aATP-BL	380,847	380,847	380,847	380,847	380,847	380,847	380,847	0
FIFS	353,615	350,811	349,190	354,186	353,483	351,195	349,136	2,119
Difference	7.15%	7.89%	8.31%	7.00%	7.19%	7.79%	8.33%	

Table 9. Number of orders denied by aATP-BL and FIFS for each scenario.

Scenario	1	2	3	4	5	6	7
aATP-BL	25	25	25	25	25	25	25
FIFS	27	28	28	27	27	27	28
Difference	2	3	3	2	2	2	3

Table 10. Orders denied drill-down by segment for scenario 4.

Segments	1	2	3	4	5	6
aATP-BL	0	0	4	0	14	7
FIFS	2	3	3	3	13	3
Difference	2	3	−1	3	−1	−4

However, it is not just the higher number of orders accepted that justifies the higher profitability achieved by the aATP-BL construction. The selection of orders from more profitable segments can be visualized in Table 10, which deploys the number of orders denied by segment using the results achieved for scenario 4.

Table 10 undisputedly shows that the aATP-BL construction prioritized the demand fulfilment for premium price markets for Canada and the USA (segments 1 and 4, respectively) and average market price segment for the Canadian marketplace (segment 2). On the contrary, the FIFS model denied orders for higher profit segments (1, 2 and 4) in detriment to lower profitable orders (segments 3, 5 and 6).

In spite of some practical simplifications (mainly the constraint of one product per order and the assumption of 100% forecast accuracy), the results may exhibit a possible boundary of profitability improvement when aATP models are pursued by commodity manufacturers. These results are aligned with the results obtained by Meyr (2009) and Quante et al. (2009a).

3.3 Impacts of increasing number of different products within an order

In order to further investigate the benefits of the proposed aATP models for practical use by lumber manufacturers, three supplementary analyses have been conducted: this section presents the examinations on increasing numbers of different products within an order ($N > 1$). As previously mentioned, more than one product per order might reduce the degree of freedom during the allocation consumption step since allocations are planned based on independent forecasts and consequently it might reduce the benefits of the aATP application. To our knowledge, the current literature has not addressed this issue so far. The following section complements these analyses by simulating the impacts of demand accuracy in the aATP proposition, considering again different average numbers of products per order. Finally, Section 3.5 analyses the impact of different levels of demand surplus (the level of demand volume above the company's capacity) in the results achieved by an aATP proposition. As mentioned earlier, aATP models intend to support profitability optimization in scenarios where demand exceeds supply. In this way, the analyses intend to verify whether the higher the demand surplus, the better the benefit of applying aATP models. The analyses are performed through different demand surplus simulation scenarios involving different numbers of products per order as well as different levels of forecast accuracy for the AP activity.

Table 11. Comparison of total number of orders per product by week for $N = 1$ and 1.95.

Products	No. of orders per delivery week $N = 1$					No. of orders per delivery week $N = 1.95$				
	W1	W2	W3	W4	Total	W1	W2	W3	W4	Total
1	5	8	8	5	26	10	16	14	10	50
2	5	8	5	8	26	10	16	10	16	52
3	8	9	6	6	29	14	18	12	12	56
Total	18	25	19	19	81	34	50	36	38	158

Table 12. Comparison of orders profile by customer segment and products in terms of number of orders and volume (truckload).

Segments	No. of orders per product $N = 1$				No. of orders per product $N = 1.95$			
	L1	L2	L3	Total	L1	L2	L3	Total
1	3	4	4	11	6	8	8	22
2	7	8	11	26	14	16	22	52
3	3	4	4	11	6	8	8	22
4	3	2	2	7	5	4	3	12
5	7	6	6	19	14	12	12	38
6	3	2	2	7	5	4	3	12
Total	26	26	29	81	50	52	56	158

Regarding the analyses that are exclusively about different numbers of products per order (forecast accuracy still at 100%), one additional customer order database was created. Using again the total of 81 orders, a new customer order database was developed with an average of 1.95 products per order. The new order database transformed the previous single product orders ($N = 1$) into 77 orders requesting two different products (0.5 TL per product) and four single product orders. The reason for this configuration lies in the requirement to keep the same order profile (Tables 5–7) in terms of number of orders per segment per week, volume of truckloads per product per week, volume of truckloads per segment per week, and finally, volume of truckloads per product per segment. Therefore, the only changes in comparison to $N = 1$ are in the number of orders per product per customer delivery week and the total number of orders per product per segment. Tables 11 and 12 demonstrate the changes provided with the new set of customer orders.

It is important to emphasize that in this new scenario, where an order can request two different products in 0.5 TL each, all models developed (GO, aATP-BL, aATP-FIFS and FIFS) are constrained by the need to source all products within an order from the same inventory location. This constraint is necessary given the limitation of this research in considering a single transportation capacity strategy (TL). Section 4 describes the opportunity to explore different transportation strategies in future research.

The average results obtained for $N = 1$ and 1.95 simulations are depicted on Tables 13 and 14, respectively. In the new sample ($N = 1.95$), the average performances of the aATP-BL construction again exceed the respective FIFS counterpart by 3.9%. It is important to observe the intense drop in performance faced by the aATP-FIFS proposition when using the order database $N = 1.95$. Such an effect is associated with the decline in

Table 13. Constructions results for $N = 1$.

$N = 1$	GO	aATP-BL	aATP-FIFS	FIFS
Average net profit ($)	380,847.00	380,847.00	380,847.00	351,659.43
Average orders denied	25.00	25.00	25.00	27.43

Table 14. Constructions results for $N = 1.95$.

$N = 1.95$	GO	aATP-BL	aATP-FIFS	FIFS
Average net profit ($)	378.835,00	356.707,00	268.712,57	342.886,43
Average orders denied	25.00	27.57	28.57	28.86

degrees of freedom for the ATP consumption caused by the need to source different products, within the same order, from a single-inventory location. First, the inventory availability is defined based on independent forecasts, and second, no availability control mechanism is in place to allow more profitable segments to have access to less profitable allocations. Therefore, the model tries to fulfil the order from a different inventory node, most probably at a higher cost, while products are sitting in lower-cost locations but reserved for less-profitable segments. The magnitude of this drop in performance achieves approximately 29% when compared with GO and 24.7% when compared with aATP-BL.

The results presented in Table 14 show the superior performance of aATP-BL construction against the FIFS model when using average products per order $N = 1.95$. However, it is important to remember that in this case the forecast accuracy applied during the aATP AP is 100%, in other words, less applicable on practical situations. The next section examines the impact on aATP-BL models of different forecast accuracy profiles.

3.4 Impacts of demand forecast accuracy

At this point, demand forecast errors are introduced into the AP step with the aim of further investigating, through exemplary situations, the boundaries for the applicability of the aATP-BL construction. In order to do so, simulations were conducted using two customer order databases with different average numbers of products per order; $N = 1$ and 1.95. For each database, five profiles of demand forecast errors were introduced during the AP step. The selection of the five demand forecast profiles was performed in partnership with sales executives of the company analysed and therefore the outcomes obtained, due to their exemplary nature, require further investigation to support more universal statements.

(1) Over-forecasted in +20%: the forecast of all products in all weeks presents an over-estimation of 20% in terms of demand volumes.
(2) Over-forecasted in +10%: the forecast of all products in all weeks presents an over-estimation of 10% in terms of demand volumes.
(3) 100% forecast accuracy: the forecast of all products in all weeks presents an 100% demand forecast accuracy, as examined in previously analyses.
(4) Under-forecasted in −10%: the forecast of all products in all weeks presents an underestimation of 10% in terms of demand volumes.
(5) Under-forecasted in −20%: the forecast of all products in all weeks presents an underestimation of 20% in terms of demand volumes.

The results of the simulations for the database $N = 1$ are presented in Table 15. In this environment, it is noticed that aATP-BL still exceeds the performance of FIFS when demand numbers are over-forecasted by 10%. However, in 20% of positive forecast errors, FIFS exceeds the results achieved by the aATP-BL proposition. This fact indicates that,

Table 15. Net profit performances (in % to GO model) for different demand forecast errors ($N = 1$).

Demand forecast error	+20%	+10%	0	−10%	−20%
aATP-BL (% of GO)	86.95%	93.04%	100.00%	95.78%	96.05%
FIFS (% of GO)	92.34%	92.34%	92.34%	92.34%	92.34%
Difference	−5.39%	0.70%	7.66%	3.44%	3.72%

in this exemplary context, positive forecast errors in approximately 10% might represent the boundary for the aATP-BL application ($N = 1$). Nevertheless, this situation must be analysed cautiously since a uniform forecast error was applied for all products and weeks along the planning horizon. Usually, demand forecast errors vary among different products and increase along the planning horizon.

On the other hand, the effects of negative demand accuracy seem to slightly influence the performance of the aATP-BL proposition. In the demand error scenarios of −10% and −20%, the aATP-BL exceeds the performance of FIFS by approximately 3.44% and 3.72%, respectively. In summary, the performance of the aATP-BL models in this environment might be explained by the fact that, in a positive forecast errors situation, volumes are reserved in excess for high margin customers that in reality will not be completely consumed due to lack of orders. In other words, in similar aATP-BL environments with a general over-forecasted situation, the net profit drops because the company set products aside for higher profitability orders (denying lower profitability orders), but these orders will not completely appear. In an analogous situation, airlines might try to sell more expensive fares for a certain flight, but end up flying with empty seats and losing potential revenue.

In contrast, when demand volumes are under-forecasted, fewer volumes are allocated for premium segments, however, the BL feature helps to partially correct this situation by allowing the acceptance of those orders (consuming from lower priority segments) when they arrive. Future research is needed to further investigate the impact of forecast error in aATP models.

The same analysis was conducted considering the order database $N = 1.95$. In this situation, the aATP-BL exceeds the performance of the FIFS construction in all scenarios of demand forecast error. On the positive values for the forecast error, the aATP-BL presents a performance of 3.65% (+10%) and 2.62% (+20%) superior to the FIFS construction. On the negative section, the outcomes are 2.88% and 2.29% better than the performance of FIFS with −10% and −20% scenarios, respectively. Table 16 displays the average performance of all scenarios analysed for the aATP-BL and FIFS constructions in comparison with the results achieved by the GO model.

Considering the scenarios with positive forecast errors (+10 and +20) the outperformance of the aATP-BL model occurs due to two main reasons: first, the drop in the FIFS performance because of the existence of more than one product in the same order with the constraint of always sourcing the order from a single location without an availability control mechanism. With this constraint in place, inventory imbalance among different

Table 16. Net profit performances (in % to GO model) for different demand forecast errors ($N = 1.95$).

Demand forecast error	+20%	+10%	0%	−10%	−20%
aATP-BL (% of GO)	91.80%	93.13%	94.16%	93.40%	92.80%
FIFS (% of GO)	90.51%	90.51%	90.51%	90.51%	90.51%
Difference	1.29%	2.62%	3.65%	2.88%	2.29%

Table 17. Net profit and number of orders denied by segment for all scenarios analysed for $N = 1.95$ and forecast error $+10\%$.

	Scenarios													
	1		2		3		4		5		6		7	
	aATP-BL	FIFS	aATP-BL	FIFS	aATP-BL	FIFS	aATP-BL	FIFS	aATP-BL	FIFS	aATP-BL	FIFS	aATP-BL	FIFS
Net profit ($)	351,009	350,195	364,470	339,746	344,503	334,264	354,040	362,493	362,715	329,220	348,272	341,481	344,615	342,806
No. of orders denied	29	28	27	29	30	30	28	26	27	31	29	29	30	29
Seg. 1	1	1	1	2	1	3	2	2	1	3	1	1	1	0
Seg. 2	3	4	1	5	4	5	3	1	3	4	4	5	4	7
Seg. 3	7	0	6	3	7	2	4	2	4	2	6	0	6	2
Seg. 4	0	4	1	4	0	3	0	5	0	3	0	5	0	3
Seg. 5	11	15	11	10	11	13	12	11	12	14	11	13	12	13
Seg. 6	7	4	7	5	7	4	7	5	7	5	7	5	7	4

products and storage locations decreases the performance of the FIFS. Second, even though the FIFS construction is capable of accepting a higher number of orders than its opponent aATP-BL, it does so by accepting lower profitable orders in detriment to higher profitable ones. Table 17 presents the net profit achieved as well as the number of orders denied for each of the seven scenarios analysed considering $N = 1.95$ and $+10\%$ forecast error. In this setting, while the average performance of aATP-BL generates a net profit of $\$352,803.00$ and denies in average 28.57 orders, the FIFS generates, on average, $\$342,886.00$ and denies 28.86 orders. However the improved performance in net profit is due to the fact that the aATP-BL construction denies a lower volume of orders for premium segments such as segments 1, 2 and 4 when compared to FIFS (except for segment 2 in the scenario 4 and segment 1 in the scenario 7).

3.5 Impacts of different levels of demand surplus

Finally, different levels of demand surplus were analysed. As previously debated, higher levels of demand surplus should support better performance of the aATP-BL model in comparison with FIFS. Three scenarios were developed using the demand and production profile presented below (Table 18) and varying the initial inventory quantity. Scenarios were again developed in partnership with sales executives of the case company.

For the original demand surplus scenarios, named here 'SUP1', the volume of initial inventory used was: 6 units for product 1 (L1), 6 units for product 2 (L2) and 12 units for product 3 (L3). For a second demand surplus scenario, named 'SUP2', the initial inventory level used was 2 units for L1, 2 units for L2 and 8 units for L3. Lastly, a third demand surplus profile, named 'SUP3', was developed. For it, the inventory data considered no initial inventory for L1 and L2 and 6 units for L3.

Table 19 presents the average result of all simulations considering two scenarios of number of products per order ($N = 1$ and 1.95), five scenarios of forecast error ($+20\%$, $+10\%$, 0%, -10% and -20%) and three scenarios of demand surplus (SUP1, SUP2 and SUP3). The results attest to the fact that the more constrained the demand volume is, the better the results achieved are by the OP that applies the aATP-BL logic in comparison with the FIFS. Even for the scenarios considering $N = 1$ and forecast error $= +20\%$, the superior performance of FIFS for SUP1 was overcome when SUP3 was applied. The main reason for this shift lies in the drop of performance of FIFS when the demand surplus increases and the capability of the aATP-BL system to prioritize more profitable orders.

Table 18. Production and demand data used for the simulations.

	Production W1	Production W2	Production W3	Production W4	Total
Product 1					
Production	3	3	3	3	12
Demand	5	8	8	5	26
product 2					
Production	2	4	2	4	12
Demand	5	8	5	8	26
product 3					
Production	2	2	2	2	8
Demand	8	9	6	6	29

Table 19. Average performance of aATP-BL and FIFS in comparison with the GO model for different settings of number of products per order, different forecast error profiles and different levels of demand surplus.

	N = 1; SUP = 1					N = 1.95; SUP = 1				
Forecast error	+20%	+10%	0%	−10%	−20%	+20%	+10%	0%	−10%	−20%
% of GO − aATP-BL	86.95	93.04	100.00	95.78	96.05	91.80	93.13	94.16	93.40	92.80
% of GO − FIFS	92.34	92.34	92.34	92.34	92.34	90.51	90.51	90.51	90.51	90.51
Difference (%)	−5.39	0.70	7.66	3.44	3.72	1.29	2.62	3.65	2.88	2.29

	N = 1; SUP = 2					N = 1.95; SUP = 2				
% of GO − aATP-BL	88.64	91.43	100.00	97.37	97.76	91.48	91.75	95.26	93.24	91.49
% of GO − FIFS	89.47	89.47	89.47	89.47	89.47	86.97	86.97	86.97	86.97	86.97
Difference (%)	−0.83	1.96	10.53	7.90	8.29	4.51	4.77	8.29	6.27	4.51

	N = 1; SUP = 3					N = 1.95; SUP = 3				
% of GO − aATP-BL	88.03	92.56	100.00	97.05	96.22	91.67	92.10	93.82	92.59	93.62
% of GO − FIFS	87.90	87.94	87.94	87.94	87.94	86.61	86.61	86.61	86.61	86.61
Difference (%)	0.10	4.61	12.06	9.11	8.28	5.06	5.50	7.22	5.98	7.01

4 Summary of findings and further research

In this paper, we explore the use of the concept of aATP with nested BLs in commodity business settings. The application of the concept intends to improve the financial outcome of the company by searching for more profitable orders in an environment where demand exceeds supply, customers present different willingness to pay and order arrival sequence is unknown. The proposed model (aATP-BL) seems efficient in general as demonstrated through the intensive exemplary simulation developed using real data from a Canadian softwood lumber manufacturer.

Considering a single product per order, the examples showed a severe drop in performance for the aATP-BL model when forecast errors are incrementally positive. The cause of the drop is associated with the fact that, in this scenario, supply volumes are allocated in excess to higher profitable market segments and will remain in inventory during allocation consumption phase since they do not become accessible for less profitable orders. On the other hand, negative forecast errors scenarios present a low impact in OP financial outcome due to the capacity of the BL feature to provide access to higher profitable segments to lower profitable allocations.

In addition, analyses were conducted in an environment where customer requests presented more than one product per order. These analyses are unique in the current literature. Using a new set of customer orders ($N = 1.95$) we were able to demonstrate that, for the analysed settings, the aATP-BL performance presents better performance than FIFS constructions even when positive forecast errors occur. In this case, the FIFS performance drops due to the constraint of having one single sourcing location for multiple products within each order. This complexity is better overcome by aATP-BL construction, which is capable of selecting orders with higher profitability.

Finally, we tested different scenarios of forecast errors and numbers of products per order for different levels of demand surplus. Here again, results demonstrated that the higher the demand surplus, the better the performance of the aATP-BL against FIFS logic. In the scenarios where demand exceeded supply, most (SUP3), improvements varied from

0.1% to 12.1% when one product per order ($N = 1$) was tested and from 5.1% to 7.2% when $N = 1.95$ was assessed.

However, further improvements must be analysed in order to advance the understanding of the real benefits and limits of this proposition in commodity business settings. First, price and demand stochasticity are very relevant in the softwood lumber industry and need to be addressed in future research. Second, more sophisticated techniques should be tested for the definition of customer willingness to pay and subsequently customer segmentation. This may create opportunities to simultaneously optimize the benefits of volumes sold through contracts and in the spot markets. In addition, different transportation strategies (i.e., less-than-truckload) and volume negotiation as well as product substitution schemes could also reduce the gap of the model to practical applications.

Another important field that requires further investigation is the existence of forecast errors in the demand allocation step. Different profiles of demand forecast errors should be analysed, and the most appropriate level of information to be used in the aATP-BL logic (e.g., product family, sub-family, SKU) should be determined.

As previously noted, the performance of BLs should be further investigated with different values for the BL consumption priority parameters ($\gamma 1$, $\gamma 2$, and $\gamma 3$). Furthermore, more frequent recalculation of BLs could possibly generate additional benefits for manufacturers.

Finally, this business problem was formulated using aATP and BLs; however, new constructions using, for example, bid prices and multi-product bundling models should also be evaluated.

The application of revenue management concepts in commodity settings has been evolving considerably in the past few years. Using real data from another softwood lumber manufacturer, Ali et al. (2014) developed a multi-level framework for demand fulfilment in a MTS environment (including an aATP model with BLs), which allowed order reassignment (changing decisions on how firm orders have to be fulfilled) in a rolling horizon approach. Further advances are needed to better integrate the concepts of revenue and supply chain management, aiming for a higher level of competitiveness for traditional manufacturers.

Disclosure statement

No potential conflict of interest was reported by the authors.

References

Alemany MME, Lario F-C, Ortiz A, Gomez F. 2013. Available-to-promise modeling for multi-plant manufacturing characterized by lack of homogeneity in the product: an illustration of a ceramic case. Appl Math Modell. 37:3380−3398.

Ali MB, Gaudreault J, D'Amours S, Carle MA. 2014. A multi-level framework for demand fulfillment in a make-to-stock environment − a case in Canadian Softwood Lumber Industry. In: 10th International Conference on Modeling, Optimization and Simulation (MOSIM 2014). Nance, France.

Ball MO, Chen CY, Zhao ZY. 2004. Available-to-promise. In: Simchi-Levi D, Wu SD, Shen ZJ, editor. Handbook of quantitative supply chain analysis: modeling in the E-business area. New York: Springer.

Chen J-H, Lin JT, Wu Y-S. 2008. Order promising rolling planning with ATP/CTP re-allocation mechanism. Indus Eng Manage Syst. 7:57–65.

Chiang DM-H, Wu AW-D. 2011. Discrete order admission ATP model with joint effect of margin and order size in a MTO environment. Int J Prod Econ. 133:761–775.

Donald WS, Maness TC, Marinescu MV. 2001. Production planning for integrated primary and secondary lumber manufacturing. Wood Fiber Sci. 33:334–344.

Eppler S, 2015. Allocating planning for demand fulfillment in make-to-stock industries – a stochastic linear programming approach [dissertation]. Darmstadt, Germany: Technical University of Darmstadt.

Forget P, D'Amours S, Frayret J-M. 2008. Multi-behavior agent model for planning in supply chains: an application to the lumber industry. Rob Comput Integr Manuf. 24:664–679.

Guerrero HH, Kern GM. 1988. How to more effectively accept and refuse orders. Prod Inv Manage J. 29:59–63.

Harris FHdeB., Pinder JP. 1995. A revenue management approach to demand management and order booking in assemble-to-order manufacturing. J Oper Manage. 13:299–309.

Johansson M. 2004. Managing the sawmill with product costs – a simulation study. Scand Forest Econ. 40:229–239.

Kalyan V. 2002. Dynamic customer value management: asset values under demand uncertainty using airline yield management and related techniques. Inform Syst Front. 4:101–109.

Kilger C, Meyr H. 2015. Demand fulfillment and ATP. In: Stadtler H, Kilger C, Meyr H, editors. Chapter 9 in Supply chain management and advanced planning – concepts, models, software and case studies. 5th ed. Heidelberg: Springer.

Lehoux N, Marier P, D'Amours S, Ouellet D, Beaulieu J. 2012. The value creation network of Canadian wood fibre. Report CIRRELT-2012-34, Centre interuniversitaire de reserche sur les résaux d'entreprise, la logistique et le transport, Québec, Canada.

Marier P, Bolduc S, Ali MB, Gaudreault J. 2014. S&OP network model commodity lumber products. Report CIRRELT-2014-25, Centre interuniversitaire de reserche sur les résaux d'entreprise, la logistique et le transport, Québec, Canada.

Meyr H. 2009. Customer segmentation, allocation planning and order promising in make-to-stock production. OR Spectrum. 31:229–256.

Pibernik R. 2005. Advanced available-to-promise: Classification, selected methods and requirements for operations and inventory management. Int J Prod Econ. 8:239–252.

Pibernik R, Yadav P. 2009. Inventory reservation and real-time order promising in a Make-to-Stock system. OR Spectrum. 31:281–307.

Quante R, Fleischmann M, Meyr H. 2009a. A stochastic dynamic programming approach to revenue management in a make-to-stock production system. ERIM Report Series Research in Management. Last accessed on July 22, 2015. Available from: http://hdl.handle.net/1765/15183.

Quante R, Meyr H, Fleischmann M. 2009b. Revenue management and demand fulfillment: matching applications, models, and software. OR Spectrum. 31:31–62.

Talluri KT, van Ryzin G. 2005. The theory and practice of revenue management. New York: Springer.

Tamura T, Fujita S. 1995. Designing customer oriented production planning system (COPPS). Int J Prod Econ. 41:377–385.

Venkatadri U, Srinivasan A, Montreuil B, Saraswat A. 2006. Optimization-based decision support for order promising in supply chain networks. Int J Prod Econ. 103:117–130.

Vogel S. 2013. Demand fulfillment in multi-stage customer hierarchies. Wiesbaden: Springer.

Yang L, Fu Y. 2008. An AATP model based on CTP for two-stage production system. Int J Bus Manag. 3:11–17.

Analysis of uncontrollable supply effects on a co-production demand-driven wood remanufacturing mill with alternative processes

Rezvan Rafiei, Mustapha Nourelfath, Jonathan Gaudreault,
Luis Antonio De Santa-Eulalia and Mathieu Bouchard

ABSTRACT

This article applies an optimization and simulation framework to identify the impacts of uncontrollable supply in the demand-driven wood remanufacturing industry. In the remanufacturing business, providing supply of raw materials in terms of quality is an uncontrollable process, especially when combined with other industry characteristics such as divergent co-production, alternative processes, make-to-order philosophy and short order cycle times. By considering a set of key performance indicators (KPIs) to measure production plan efficiency, our framework uses a periodic re-planning strategy based on a rolling horizon. Then, the source of lumber is perturbed, leading to different supply scenarios. Simulations are conducted using a mixture experimental design approach to identify the safety threshold of supply changes in the scenarios regarding the selected KPIs. This threshold of all KPIs leads us to an overall optimum region which includes a series of desirable scenarios. The proposed framework allows planner to understand the impact of supply policies to deal with uncertainties.

1. Introduction

Most industries are forced to seek new ways to increase profitability and reduce costs in order to compete in current economic conditions. Like other industries, the softwood forest products industry is affected by international competition. The arrival of products from countries with low-cost labour supply has made price-based competition very difficult. Companies thus have to decrease their costs and improve the value of offered products (Lihra 2007). In this context, the primary processing plants (sawmills) or their holding companies in the forest industry are motivated to develop value-added products resulting from secondary and tertiary processing plants, known as wood remanufacturing mills. Both primary and secondary processing of wood have become an important source of employment and income generation (Schmincke 1995). Supporting secondary industries is thus recognized as an important parameter for the development of the forestry sector. This requires improving manufacturing processes to boost efficiency and ensure return on capital investment.

Remanufacturing wood mills produce many types of end-products at the same time in a production run following a *divergent co-production* logic. This *co-production* phenomenon cannot be avoided due to an inherent process characteristic in the manufacturing system. In this type of system, manufacturers may face shortages of products that are more demanded by customers and surplus of products that are unwanted, mainly due to the uncertainty of the co-production process. Moreover, a given component can be produced using different *alternative processes* (recipes), consuming different types of lumber and producing different sets of co-products through a divergent manufacturing process. Furthermore, the market is dynamic with a wide range of products and *changing demands* which are received on a short-term basis. According to the current practice in the market, production needs to be demand-driven and planned according to a make-to-order philosophy (while co-products are inevitably stocked).

Another point is that the softwood lumber is classified based on grade, dimension, form, species and condition or degree of seasoning at time by the National Lumber Grades Authority (NLGA) standards in Canada. Despite classification of lumber in sawmills, lumber is reclassified in the remanufacturing mill as NLGA standards do not take into account certain appearance characteristics that have great importance for the secondary plants. In such plants, unusable areas (imperfects) spread throughout the lumber are a major factor in the grading. The number of knots and their location, splits, crooks, or flaws are the determinant for grading. But, as wood is a biological product, the exact production yield cannot be determined a priori and depends on supply quality.

Moreover, sawmill products are generally forwarded to retailers and distribution centres (e.g. in the United States) with long-term contracts. Under these contracts, their best products are often presold to these customers, thus sawmills are not usually able to offer their best lumber to local secondary plants. Some remanufacturers consider switching to different suppliers from outside their local regions to deal with this issue. In this case, delivery problems and transportation costs are serious impediments for remanufacturers, even if the new suppliers offer guarantees related to better supply quality. Another point is that even if sawmills sell the best products or required raw materials quality to the secondary plants, it may not be economically reasonable for secondary plants. Nevertheless, some remanufacturers tend to purchase higher quality lumber to enable it to cover all quality types of orders. The reason is that it is always possible to produce lower-quality end-products from superior quality raw materials; however, this industrial practice is expensive.

The case under study is a manufacturer of bed frame components that need long or wide lumber to be cut from high quality class, as well as small items that can be produced from lower-quality lumber. In general, sawmills, usually owned by the same holding company, are less concerned with this local secondary plant but more concerned with supplier reputation in the global market. As a result, the ability to obtain wood of a desired quality is seen as 'uncontrollable' by the remanufacturer, thus procurement is a very challenging activity for the company. Under these circumstances, the company can take advantage of cost minimization by purchasing a mix of high- and low-quality lumber through finding the best match between lumber grade and lumber quantity requirements order. However, identifying desirable mixes of qualities of supply is a complex task for planners especially for service-sensitive companies, like the case under study.

In this context, this work analyzes the effect of supply uncontrollability on the production planning based on the selected key performance indicators (KPIs) in order to provide managerial insight to improve supply chain practice. A wood remanufacturing mill in Eastern Canada is considered as a case study.

The research questions answered in this paper are:

1. What are the impacts of not having control over the supply quality in the wood remanufacturing industry?
2. Is there any safety threshold of supply changes for the industrial case under study?

As demonstrated in the next section, the scientific literature has not approached these research questions before.

The remainder of the paper is organized as follows. Section 2 provides a brief literature review of some related work and paper contributions. After that, we present the analysis methodology and the experimental results in Sections 3 and 4, respectively. Section 5 concludes the paper.

2. Literature review and paper contributions

2.1. Literature review

In this section, we first show how different industries deal with co-production systems, uncontrollable supply and changing demands in the context of production planning. Then, the approaches used in the forest industry to solve similar production planning problems are discussed.

Two recent research papers in the semiconductor industry were proposed by Han et al. (2012) and Wang (2009). Han et al. (2012) considered yield variability in a co-production system. In that case, the end-products quality was uncertain and graded into several quality levels. A stochastic dynamic program with profit maximization was used to model the production problem. Wang (2009) studied a decentralized semiconductor supply chain comprising a single manufacturer and a single distributor for a short life-cycle product with random yield and uncertain demand.

Boyabatli et al. (2011) analyzed the optimal procurement, processing and production decisions of a meat processing company. They considered two beef products in a coproduction context where the premium product allowed downward substituting to meet market demand. For this purpose, the two-stage stochastic approach was applied to model the problem.

In the context of reverse logistics, some recent research has been proposed as well. Shi et al. (2011) provided a production plan for a multi-product closed-loop system. Two different channels of supplying products have been considered by Shi et al. (2011) including brand-new products and remanufactured products. For returned products, demands were uncertain and price-sensitive. Their objective was to maximize the profit through producing brand-new products and determining quantity of remanufactured products as well as the acquisition prices of returned products. The proposed solution was based on a

Lagrangian relaxation. In addition, Mukhopadhyay and Ma (2009) studied a hybrid production system including returned and new products as inputs. Using a two-stage stochastic analysis, the authors determined the optimal procurement and production quantity for the firm. Also, Denizel et al. (2010) developed a production plan in a remanufacturing environment with uncertain quality and deterministic demand in a multi-period setting, using a stochastic programming model.

Although many papers have focused on production planning optimization in the forest supply chain during recent years (see, e.g. D'Amours et al. (2008) for a literature review), few authors have worked on the specific problem of softwood remanufacturing production planning. We provide here a brief review of some of the papers in the forest industry.

Gaudreault et al. (2010) investigated coordination mechanisms between different production units in the lumber supply chain. The supply and demand interactions among involved unities were defined through different mechanisms via deterministic models. Moreover, Feng et al. (2010) presented rolling horizon simulation models and evaluated the performance of integrated sales and operation planning in the wood-based panel industry. In this study, the wood supply was affected by seasonal harvesting operations and long replenishment lead time and the raw material supply was evaluated based on the purchased quantity. Kazemi Zanjani et al. (2011) proposed a multi-period, multi-product sawmill production planning problem where the yields of processes were random variables due to non-homogeneous quality of raw materials (logs). Because of natural resource characteristics, the quantities of lumber that can be produced by each cutting pattern (processes yields) are random variables. In that regard, they proposed the optimal combination of log classes and cutting patterns that best fitted against lumber demand. Furthermore, Santa-Eulalia et al. (2011) investigated the variability of supply, manufacturing and demand in the softwood lumber industry using a simulation platform combining agent-based methods with optimization models. More recently, Rafiei et al. (2014, 2015) proposed periodic re-planning models to deal with uncertain demand in a wood remanufacturing mill with above-mentioned characteristics while supply of raw materials in terms of quality and quantity is infinite.

Many research papers have focused on improving lumber yield through finding optimum cut-up patterns of a board, which is a typical cutting stock problem (Rönnqvist & Astrand 1998; Ostermark 1999; Buehlmann et al. 2010; Kazemi Zanjani et al. 2011). Although the increased yield from lumber is a way to minimize raw materials costs, another issue for a wood products manufacturer (Buehlmann et al. 2011) is reducing costs through optimization of the lumber purchases, especially through price management of the different lumber quality classes. This issue is expressed as least-cost lumber quality-mix problems and is of high economic interest to industry. Significant research has been conducted to determine yields of least-cost grade mixes in this industry through specialized software packages developed over time (Harding & Steele 1997; Buck et al. 2010). For example, Buehlmann and Thomas (2001) proposed a search algorithm in simulation models to find the optimal cutting pattern in the secondary wood industry. In a more recent study Buehlmann et al. (2011) proposed a new least-cost lumber quality-mix model with lower production costs in the secondary wood industry. Hoff et al. (1997) also reported difficulties with raw material supply (grade and price) in the US manufacturers. Moreover, Kozak et al. (2003) had a survey conducted in British Columbia, Canada, about the impediments to wood supply relationships between secondary and primary manufacturers. They propounded that the lumber remanufacturing industry is a complex case and the majority of secondary manufacturers were experiencing lumber procurement problems.

2.2. *Research gap and paper contribution*

The above literature review confirms that although studying the supply variations under production planning has attracted attention in several industry sectors, it is still an emerging research field with several open research questions. Specifically, in the forest supply chain, there have been opportunities in the area of production planning optimization of the primary processing plants (sawmills) regarding co-production system, process yield, heterogeneous log quality and dynamic market. Nevertheless, the inherent characteristics of the softwood remanufacturing industry (e.g. imperfect quality raw material, supply variation, alternative process, make-to-order system and short customer lead time) differentiate it from the primary processing plants, making this sector even more complex to manage. In general, research in the secondary plants (remanufacturing ones) focuses mostly on improving lumber yield (e.g. Harding & Steele 1997; Buehlmann & Thomas 2001; Buck et al. 2010; Buehlmann et al. 2011), while some others recognize lumber procurement problems (e.g. Hoff et al. 1997; Kazak et al. 2003). Finally, it is important to say that some recent studies deal with the production planning problem in the remanufacturing sector, but they consider unlimited supply (Rafiei et al. 2014, 2015), which is a simplification of the reality. This leads us to believe that companies are not aware of the magnitude of the impact that supply variations may have on the supply chain planning performance, and the scientific literature does not provide solutions for this issue.

To contribute to reducing this research gap, in this study, we apply the proposed optimization/simulation framework in Rafiei et al. (2014) to assess the performance of the current industrial practice, in order to ultimately guide planners in improving their production plans. The main contribution of this framework is its capacity to 'emulate' the current industrial practice, by generating production plans that have the ability to react to changing demands using a re-planning approach with periodic policy. We adapted the work of Rafiei et al. (2014) in terms of managing supply changes, also providing new capabilities for scenario generation and analysis. We first compute a production plan with a presumably 'ideal' supply. Here, the word 'ideal' means that the supply quality required by the generated production plan is always available. Since such ideal supply rarely happens, this plan represents only a reference to evaluate the performance deviation under more realistic supply. To take into account the uncontrollability of supply quality, such ideal supply is perturbed and used as a constraint in the production planning mathematical model in order to measure by simulation the impacts on the mill performance. The perturbation mechanisms used are inspired from real supply uncontrollability encountered in the industrial case. For a realistic analysis considering companies' and customers' viewpoints, we evaluated different KPIs, such as customer order fulfilment, total cost and inventory levels. Through this decision framework, decision-makers can assess the impact on the supply chain performance in terms of KPIs. Moreover, it is possible to determine a safety threshold for the uncontrollable supply in the scenarios regarding the selected KPIs. The threshold of all KPIs leads to an overall optimum region comprising a series of desirable scenarios. According to this safety threshold, decision-makers can purchase the combination of lumber qualities, remarkably decreasing raw materials costs and backorder quantities. As a result, the analysis may serve not only as a diagnosis tool to quantify and evaluate how good (or bad) the current industrial practice is, but it can also be used as a platform to optimize, simulate and analyze alternative production planning policies prior to their implementation in the wood remanufacturing area. To the best of our knowledge, research published so far has not considered the wood remanufacturing industrial problem addressed in this study.

3. Methodology

3.1. General steps

3.1.1. Defining the performance indicators
The first step of our general methodology consists in defining the performance criteria to be used in the measure of the impacts of the supply uncontrollability. In our case, the used KPIs are the backorder quantity, the cost and the inventory level. These indicators have been defined with our industrial partner. Inventory level and production costs refer to the company viewpoint, while the backorder represents the clients' viewpoint. Note however that knowing which supply chain KPIs are important to monitor can be difficult.

3.1.2. Measuring the current performance
To evaluate the current KPIs, we need to define how the production planning process is performed. In the studied mill, there is a spreadsheet tool (using Excel) that allows planners to observe the results of using different alternative processes. However, planners' intuition and experience play an important role in making a decision with this tool. To optimize production planning while handling the changes in the demand, an optimization/simulation framework was developed for this industrial case by Rafiei et al. (2014). Based on mathematical programming and simulation, this framework has the ability to react to changing demands using a re-planning approach with periodic policy. Note that we assume, without generality loss, that the production plan generated by the optimization/simulation framework corresponds to the plan generated by the planner. As this framework was not able to deal with supply quality variations, we developed an improved version providing additional capabilities for scenario generation and analysis in terms of managing uncontrollable supply. The use of the optimization/simulation framework allows for a quick automatic generation and analysis of the planning solutions. In addition, it allows for superior performance, since optimization is being performed.

Before presenting the optimization/simulation framework (in Section 3.3), let us briefly explain the remaining steps of the methodology.

3.1.3. Computing the production plan under full supply availability
The previous framework indicated the ability to react to changing demands using a re-planning approach with periodic policy without considering supply variations. We use this framework to compute a production plan with a presumably ideal supply, i.e. assuming that the supply quality required is always available. This plan will be used as a reference to evaluate the performance deviation under more realistic (uncontrollable) supply in this study.

In addition, to obtain the ideal supply, we use an optimization model that makes efforts to meet customer demand using the entire capacity. In that regard, the model objective function is the backorder quantity minimization. Therefore, here the ideal supply originates from the objective of keeping the customer satisfied.

3.1.4. Measuring the impacts

According to the current practice of the case under study, we considered three quality classes for end-products: high-quality (HQ), mid-quality (MQ) and low-quality (LQ). As we already discussed, the wood remanufacturers usually do not have issues providing the total quantity of the required supply; however, these materials are supplied with different and uncertain proportions of quality classes. To deal with this situation, we generate different scenarios representing supply quality variations by taking inspiration from the case under study. The objective of generating these scenarios is to measure the impacts of the changes in the supply on the KPIs via the simulation/optimization framework.

3.2. Developed optimization/simulation framework

In this section, at first we present the mathematical model and define KPIs. Next, we have a brief review on the periodic re-planning along with rolling horizon planning. Finally, the developed decision framework is presented.

3.2.1. The mathematical model

A. *Definitions and notations.* The co-production system produces a family F including f products simultaneously from an input i of a set of raw materials I through an alternative process defined by a recipe r from a set of recipes R. From now on, we use the word alternative process to refer to a recipe.

As reported before, three quality classes are usually distinguished in this case: HQ, MQ and LQ. To produce an end-product of a given quality class, it is required to have a supply with at least the same quality. The R_{HQ} is, thus, denoted the set of recipes producing HQ end-products, the R_{MQ} is denoted the set of recipes producing MQ end-products, and finally, the R_{LQ} is expressed as the set of recipes producing LQ end-products. Besides that, it is always possible to produce lower-quality end-products from superior quality raw material. In fact, the practice in this industry consists in using higher-quality raw materials to produce co-products used for *demand downward substitution*, which means that end-products of superior quality replace demands of inferior quality end-products. The R_{AF} is denoted the set of recipes applied to produce end-products of inferior quality from superior quality raw materials. Therefore, the set of recipes is divided into four sub-sets: $R = R_{HQ} \cup R_{MQ} \cup R_{LQ} \cup R_{AF}$.

Furthermore, the production proportion for each product $f \in F$ by recipe $r \in R$ is denoted by α_r^f. Note that the sum of products proportions is equal to $(1 - \beta)$, where parameter β shows the proportion of imperfection for the given raw material. We consider a planning horizon including T periods. All periods have the same fixed length. For each product, a demand is to be satisfied at the end of period t ($t = 1, 2, \ldots, T$).

The following notations are also used:

d_t^f Demand of product f by the end of period t

s_{it} Supply of raw material i at the beginning of period t

δ_r Required capacity time for recipe r

c_t Available capacity time in period t

φ_{ir} Unit of raw material i consumed by recipe r

i_{min}^f Safety stock level of product f

p_t^r Variable production cost incurred for every recipe r produced in period t

h_t^f Holding cost charged for each product f carried in inventory in period t

The decision variables are:

X_{rt} Number of times that recipe r is used in period t

I_{it} Inventory size of raw material i by the end of period t

I_{ft} Inventory size of the end-product f by the end of period t

B_{ft} Backorder size of the end-product f by the end of period t

B. Optimization objectives. Each KPI corresponds to an objective function in the optimization model. The first KPI is measured by the backorder quantity:

$$KPI1 = \sum_{t \in T} \sum_{f \in F} B_{ft} \tag{1}$$

The second KPI is measured according to the sum of the variable production cost and the inventory holding cost:

$$KPI2 = \sum_{t \in T} \left(\sum_{r \in R} \sum_{f \in F} p_t^r \alpha_r^f X_{rt} + \sum_{f \in F} h_t^f I_{ft} \right) \tag{2}$$

It should be noted that this KPI includes inventory holding costs. As the inventory holding cost is relatively very low in this case, the total value of the second part in Equation (2) is negligible in comparison with the first part (production cost). However, high inventory quantities of (undesired) co-products are always observed and it is crucial to handle the level of inventory in such mills. Hence, to take into account the inventory quantity, the third KPI is defined as follows:

$$KPI3 = \sum_{t \in T} \sum_{f \in F} I_{ft} \tag{3}$$

Note that we consider the inventory level of products (I_{ft}) in the second part of Equation (2) ($h_t^f I_{ft}$) as well as in Equation (3). This consideration may seem a conflict at first glance; however, the low cost of holding inventory on this case causes the $h_t^f I_{ft}$ to take small values and does not have a significant effect on the total value of KPI2.

The model is mathematically formulated as:

$$\text{Minimize} \quad \sum_{t \in T} \sum_{f \in F} B_{ft} \qquad (4)$$

Subject to

Flow equilibrium constraint for raw material

$$I_{it} - I_{i(t-1)} = s_{it} - \sum_{r \in R} \varphi_{it} X_{rt} \quad \forall t \in T; \forall i \in I; \forall r \in R \qquad (5)$$

Production capacity constraint

$$\sum_{r \in R} \delta_r X_{rt} = c_t \quad \forall t \in T; \forall r \in R \qquad (6)$$

Products proportion constraint

$$\sum_{f \in F} \alpha_r^f + \beta = 1 \quad \forall r \in R \qquad (7)$$

Flow equilibrium constraint for end-products

$$I_{ft} - I_{f(t-1)} - B_{ft} + B_{f(t-1)} - \sum_{r \in R} \alpha_r^f X_{rt} = d_t^f \quad \forall t \in T; \ \forall f \in F; \forall r \in R \qquad (8)$$

Safety stock constraint

$$i_{\min}^f \leq I_{ft} \quad \forall t \in T; \ \forall f \in F \qquad (9)$$

Non-negativity of all variables

$$X_{rt} \geq 0, I_{it} \geq 0, I_{ft} \geq 0, B_{ft} \geq 0 \quad \forall t \in T; \forall i \in I; \forall f \in F; \forall r \in R \qquad (10)$$

The objective function (Equation (4)) minimizes the sum of backorder quantities at each period. Constraint (5) guarantees that the inventory of raw materials is balanced in each period. Constraint (6) ensures that the total production time is equal to the available production capacity in each period, which means the model is forced to use the entire production capacity of each period. The entire production capacity is used according to constraint (6) in this study because it is the current practice of the case under study. Constraint (7) requires that total production proportions for a unit do not exceed one. Constraint (8) is introduced to ensure flow conservation for each end-product. Constraint (9) forces the model to keep the inventory level of end-products at least at the level of safety stock. The last set of constraints (10) is the non-negativity requirement constraints.

3.2.2. Periodic re-planning as a production planning approach

The periodic planning approach along with rolling horizon is used in this study. Thus, let us briefly explain a periodic re-planning approach along with rolling horizon planning

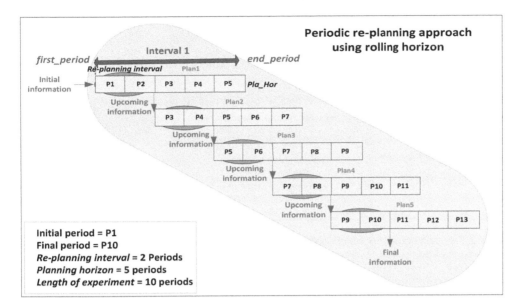

Figure 1. Illustration of periodic re-planning approach using rolling horizon.

used in our industrial context. Two important factors have to be considered in a periodic re-planning approach: the control level and planning horizon. The control level corresponds to the periodic re-planning interval, and it is denoted by Rep_Int. The second factor, the planning horizon, represents the amount of future time that will be considered when preparing a plan; it is denoted by Pla_Hor. Typically in a periodic re-planning approach, the plan is decomposed into a series of static plans at regular intervals and plans are not revised until the next interval arrives. Under such conditions, the plan is computed for a horizon in order to take into account deliveries that are due for a planning horizon (Pla_Hor), but only the re-planning interval (Rep_Int) of the plan is executed. For more details about periodic re-planning and its literature review, the reader is referred for example to Rafiei et al. (2014).

Figure 1 illustrates the re-planning approach by periodic policy, considering an example where the length of experiments is equal to 10 periods of planning activity. Periods 1, 2, ... are denoted by P1, P2, ... (respectively). In this example, the periodic re-planning interval Rep_Int is equal to two periods. The planning horizon Pla_Hor is equal to five periods. This means that the plan is revised every two periods and for a total of five times. In fact, using the initial information available at the first (initial) period P1, the first plan (Plan 1) is produced for the first interval (Interval 1, from P1 to P5). That is, the initial information, such as initial inventory of raw materials and finished products, will be the input data of the first interval. Plan 1 makes a decision in the first interval based on the relevant available information for five periods. However, Plan 1 is implemented only for periods P1 and P2. At the end of P2, as new (upcoming) information is available, the plan is then revised to determine Plan 2 for the next five periods (from P3 to P7). Indeed, the output of Plan 1 also, such as inventory of raw materials, backorders and finished products quantities at the end of period 2, will be the input of the second interval. Again, Plan

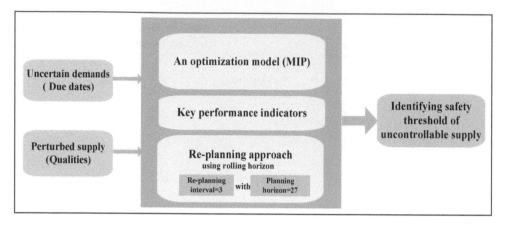

Figure 2. Decision framework for uncontrollable supply safety threshold identification.

2 is implemented only for P3 and P4. As shown in Figure 1, this process continues until the final plan (Plan 5) is obtained for the final period considered (P10).

3.3. The resulting decision framework

The last step of the proposed methodology consists in the experimentation of the decision framework. Let us first sketch this framework before presenting the experiments in the next section. Figure 2 describes the main components of the resulting optimization/simulation platform. The inputs consist of uncertain demands defined by varying of due dates and a set of scenarios consisting of perturbed supplies (by changing the percentage of supply qualities). Based on the selected KPIs, the platform uses the planning policy which consists of the triple of a planning approach, a re-planning interval and a planning horizon, to distinguish *safety thresholds* of uncontrollable supply in the case under study.

The *safety threshold* of the uncontrollable supply refers to the region at which the value changes of each KPI reaches the admitted range while the supply is changing. The safety threshold is crossed when the value changes of KPIs become threatening to the benefit of company.

4. Simulation experiments

4.1. Experimental environment

Data from a wood remanufacturing mill located in Quebec (Canada) is used to analyze the impact of supply changes in the production planning. To achieve this goal, the performance of different scenarios is compared with the ideal supply based on the considered criteria. The model is implemented in ILOG OPL STUDIO version 6.3 and is solved by CPLEX 12.4.

The case involves a total of 107 products. Orders were the real data for a 27-day planning horizon, while the mill unit worked five business days every week. Each day includes three shifts. One shift was considered as a period. Therefore without considering weekends and holidays included in a 27-day planning horizon, a total of 18 business days of 3

Table 1. Defined ranges for KPIs.

KPI	Results admissible ranges					
	Extremely Desirable	Desirable	Acceptable	Undesirable	Unacceptable	Extremely Unacceptable
Backorder QTY	Less than 450	451–1000	1001–2000	2001–5000	5001–6000	More than 6000
Costs value ($)	500,000–600,000	600,000–615,000	615,001–626,000	626,001–650,000	650,001–660,000	More than 660,000
Inventory QTY	168,000–169,000	169,001–170,000	170,001–172,000	172,001–174,500	174,501–176,500	More than 176,500

shifts each resulted in 54 periods. The number of orders was 565 orders where an order includes one product.

In this mill the demands were often responded to within two business days after the order arrival. Note however that no sample data were available for the exact due dates values. Because of this, triangular probability distributions have been chosen to randomly generate such due dates. As we know from our partner that the due dates are between one period and eight periods, the minimum and the maximum values of the used triangular distribution are one period and eight periods. Applying a symmetric triangular distribution, the modal value is considered equal to four periods. Moreover we consider a level of re-planning period length 3 corresponds to revising the plan every three periods (every day). We also considered a level of planning horizon (9 business days or 27 periods) to perform scenarios.

The KPIs include the backorder level, cost and inventory level. The backorder is of great relevance for measuring customer satisfaction while the cost is related to the company's point of view. Finally, even if the inventory holding cost is relatively low in this industry, high inventory quantities of (undesired) co-products are always observed.

We suppose that the impact of supply variations on the value of KPIs can be defined in several ranges; *Extremely Desirable, Desirable, Acceptable, Undesirable, Unacceptable* and *Extremely Unacceptable* ranges. The region with the highest desirability is called *Extremely Desirable. Desirable* range represents the second level of desirability, and acceptable results lie in *Acceptable* range. The scenarios with lower desirability fall in the *Undesirable* range and the next lower desirability range is called *Unacceptable*. And finally, the *Extremely Unacceptable* range has the lowest desirability. For each KPI the ranges are defined in Table 1. The ranges are restricted by the company's point of view regarding the preservation of customer satisfaction, reducing costs and stocking policy. In general, determining the desirability of supply chain KPI depends on business needs.

To obtain confidence intervals of 90% as reasonable levels of confidence according to Law and Kelton (2007), the number of replications in simulation model is determined using two different methods proposed by Itami et al. (2005). The results show that more than 35 replications are needed to achieve the defined relative precision. Thus, 40 replications were then performed for each scenario. As we are considering 40 replications for 38 scenarios, $(40 \times 38 =)$ 1520 simulation runs should be performed.

4.2. Experimental design

To design experiments, we consider a mixture experiment design that is a type of factorial design with dependent factors, an approach applicable in similar divergent industries like pharmaceutical, food, steel, glass, paper and lumber. These factors are the components or

ingredients of a mixture, and the response is a function of the proportions of each component and the percentages of the components must add up to 100% (Myers 1999).

In the lumber remanufacturing industry Buehlmann and Thomas (2001) and Buehlmann
et al. (2011) apply a mixture design. They explain that the amount of each individual lumber
quality including the raw material supply is between 0% and 100% of the total lumber volume used. They also consider five quality classes of lumber. As already mentioned, in our
case the raw material supply quantity is fixed; however, the proportions of each individual
lumber quality are changed. We consider three quality classes of lumber and each one is
treated as an individual factor. Interactions (first and second order) among factors are relevant, since this interaction has a straightforward effect on the selection of the alternative processes, and the alternative process selection also has serious effects on KPIs. Hence, we
investigate the impact of trade-offs among factors on KPIs through a mixture experiment
design.

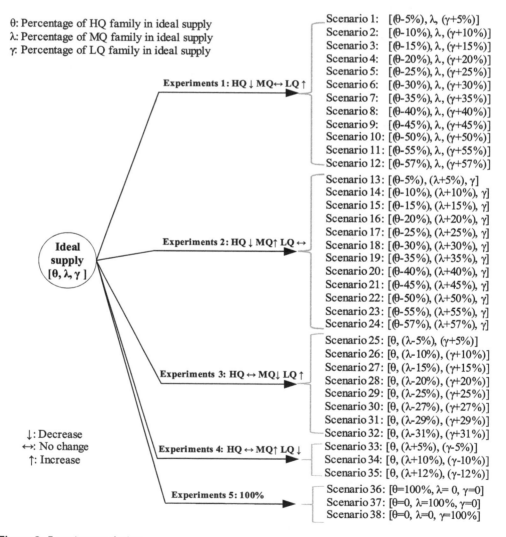

Figure 3. Experiments design.

The different raw materials consumed by the mill can be aggregated into different products classes according to the quality: HQ, MQ and LQ. To simplify this, suppose that the needs of the mill (according to the production plan) in terms of percentage value for HQ class is θ, for MQ class is λ and for LQ class is γ. The factors θ, λ and γ are related, and varying the levels of one inevitably produces a corresponding variation in the levels of the other.

As said before, we keep the ideal supply components as starting point and decrements of 5% (or less) of the ideal supply are used to create the new supply combinations. This leads to the experimental plan reported in Figure 3. We generate five experiment groups and gradually change the percentages of two components in the ideal supply at the same time. Each group consists of a certain number of scenarios. Suppose that scenarios are a set of $\{S_I, S_1, S_2, S_3, \ldots, S_{38}\}$ where the ideal supply is denoted by S_I and Scenario 1 is denoted by S_1 and so on.

(1) Experiments 1: Suppose that the proportion of λ has no change. In this situation, the percentage of θ gradually decreases down to zero, while the percentage of γ gradually increases by the same amount (Scenarios 1−12).

(2) Experiments 2: Suppose that the proportion of γ has no change. The percentage of θ gradually decreases down to zero, while the percentage of λ gradually increases by the same amount (Scenarios 13−24).

(3) Experiments 3: Suppose that the proportion of θ has no change. The percentage of λ gradually decreases down to zero, while the percentage of γ gradually increases by the same amount (Scenarios 25−32).

(4) Experiments 4: Suppose that the proportion of θ has no change. The percentage of γ gradually decreases down to zero, while the percentage of λ gradually increases by the same amount (Scenarios 33−35).

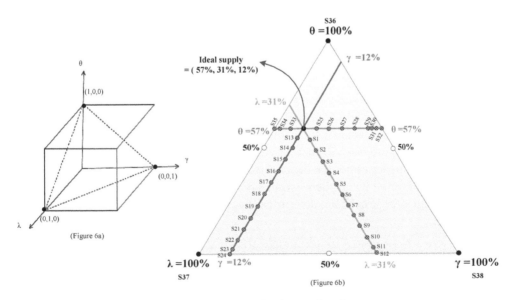

Figure 4. (a,b) Thirty-eight mixes of the three-factor lumber quality classes.

(5) Experiments 5: Based on the assumption that all of the supplied raw materials belong to only one quality class, Scenarios 36−38 are generated.

The geometric description of the three factors is a three-dimensional (3D) cube (Figure 4(a)). The factor space containing these factors consists of all points on or inside the boundaries of an equilateral triangle (Figure 4(b)) (Cornell 2011). Each of the three sides of the triangle represents a factor. In Figure 4(b), the black point shows the coordinate of the ideal supply. To obtain the coordinate of the ideal supply, first we run the optimization model for different re-planning configurations, namely, four re-planning intervals and five planning horizons based on the case limitations. In fact, the values of Rep_Int were from 1 to 4, where 1 corresponds to revising the plan every period and Pla_Hor had four levels 6, 9, 12, 15 and 27. The reason for using those specific re-planning configurations is that they were chosen by Rafiei et al. (2014) as good performance configurations in the case under study. For each plan, we compute the required supplies according to the three considered classes of quality. The percentage of sum of supplies in each class of quality leads us to the 57% HQ, 31% MQ and 12% LQ, which is considered as the ideal supply proportions ($\theta = 57\%, \lambda = 31\%, \gamma = 12\%$). In Figure 4(a), the grey points represent the coordinates of considered mixes of lumber qualities. The points in Figure 4(b) (scenarios) are defined based on possible changes which occur in the real world of the case.

In Figure 4(b), the green line (with $\lambda = 31\%$) shows the 12 scenarios (Scenarios 1−12) generated in Experiments 1. The red line (with $\gamma = 12\%$) represents the 12 scenarios (Scenarios 13−24) generated in Experiments 2. The right-hand side of the blue line (with $\theta = 57\%$) shows the eight scenarios (Scenarios 25−32) generated in Experiments 3. The left-hand side of the blue line (with $\theta = 57\%$) shows the three scenarios (Scenarios 33−35) generated in Experiments 4. Finally the bold points are representative of scenarios in Experiments 5($\theta = 100\%$, $\lambda = 100\%$ and $\gamma = 100\%$).

4.3. Experimental results

In the previous section, we show how the perturbed supply is obtained via defined scenarios. We have already defined an optimization model with three KPIs and a periodic re-planning approach along with rolling horizon planning. According to the framework (Figure 2), we should experiment on the production plan of each scenario and compare it with the ideal supply scenario. The goal of the experiments is to show that some scenarios should be unacceptable, whereas the company can deal with supply changes in some scenarios. For this purpose, we will first survey the impact of uncontrollable supply in each KPI to show the impact of not having control over supply by planners. Then, a threshold will be computed for all KPIs where estimated points meet the criteria requirements. This threshold, called safety threshold, determines degree of flexibility for planners in suppliers' negotiations.

4.3.1. Backorder level

Minimum service variability is very essential in service-sensitive manufacturing environments like wood remanufacturing mills. For this purpose, this subsection seeks scenarios with the minimum backorder size.

We studied the possible models (linear, quadratic, special cubic, cubic, special quartic and quartic) to show the response surface of the backorder using Design-Expert software 8.0.7.1 (Stat-Ease Inc., Minneapolis, MN). The analysis of variance (ANOVA) among these models shows that the special quartic model has a large predicted R-squared (0.9871) and the smallest predicted residual sum of squares (PRESS) statistic as well as root mean square error (MSE) less than others. Cornell (2011) states that a special quartic is especially useful for detecting curvature of the surface in the interior of the triangle. As the considered scenarios often lie in the interior of the triangle, as well as ANOVA results, the special quartic model is chosen to fit the response surface in this KPI.

Equation (11) is the special quartic model in terms of observations, where y is the backorder quantity and x_1, x_2 and x_3 are the proportions of each lumber grade; β_0 is the constant; β_1, β_2 and β_3 are the coefficients of linear terms; β_{12}, β_{13} and β_{23} are the coefficients of the two-term interactions; β_{1123}, β_{1223} and β_{1233} are the coefficients of the special three-term interactions. The difference between observation y and the fitted model $\hat{y}$ (in Equation (12)) is a residual, say $\varepsilon = y - \hat{y}$.

$$\begin{aligned} y = {} & \beta_0 + \beta_1\, x_1 + \beta_2\, x_2 + \beta_3 x_3 + \beta_{12} x_1\, x_2 + \beta_{13}\, x_1\, x_3 + \beta_{23} x_2\, x_3 + \beta_{1123} x_1^2\, x_2\, x_3 \\ & + \beta_{1223} x_1\, x_2^2 x_3 + \beta_{1233}\ x_1\, x_2 x_3^2\ + \varepsilon \end{aligned} \tag{11}$$

By doing the experiments, the coefficients of the equation are estimated for the backorder quantity (Equation (12)).

$$\begin{aligned} \hat{y} = {} & \hat{\beta}_1\, x_1 + \hat{\beta}_2\, x_2 + \hat{\beta}_3 x_3 + \hat{\beta}_{12}\, x_1\, x_2 + \hat{\beta}_{13} x_1\, x_3 + \hat{\beta}_{23} x_2\, x_3 + \hat{\beta}_{1123} x_1^2\, x_2\, x_3 \\ & + \hat{\beta}_{1223}\, x_1\, x_2^2 x_3 + \hat{\beta}_{1233}\ x_1\, x_2 x_3^2 \end{aligned} \tag{12}$$

Note that the constant (β_0) is not present in Equation (12); because the constant exists while all the factors x_1, x_2 and x_3 have level 0. However, in the mixture design, this condition is not possible, because the sum of all factors must be 1. The estimated model for the backorder response is expressed in Equation (13).

$$\begin{aligned} \text{Estimated backorder function} = {} & 1822.44\, x_1 + 5450.13\, x_2 + 7758.22\, x_3 - 13626.97\, x_1\, x_2 \\ & - 14470.87\, x_1\, x_3 - 8184.99 x_2\, x_3 + 69325.81 x_1^2\, x_2\, x_3 \\ & - 91738.52\ x_1\, x_2^2 x_3 - 70877.37\ x_1\, x_2 x_3^2 \end{aligned} \tag{13}$$

As P-value for coefficients of the two-term and three-term interactions is less than 0.0001 in this model, these coefficients are significant and cannot be removed from the model. Note that the coefficients of the linear terms correspond to the response obtained with the pure factors. The coefficients of the two-term interactions show the synergic effect of the two factors. Moreover the coefficients of the three-term interaction indicate the synergic effect of the three components. In the example described above, the experiment with pure x_1 ($x_2 = 0, x_3 = 0$) gave a response of 1822.4, while the experiment with pure x_3 ($x_1 = 0, x_2 = 0$) gave a response of 7758.22. If no synergic effect were present, the mixture made $x_1 = 0.57$ and $x_3 = 0.43$ ($x_2 = 0$) would give a response of 4374.80. The response of this mixture is instead 827.99 (3546.8 units less), meaning that a significant

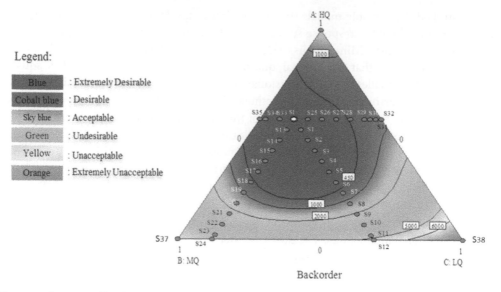

Figure 5. Contours of backorder level.

negative synergic effect is present. This can be found in the term $x_1 x_3$, whose coefficient is -14470.87.

By looking at the coefficients of Equation (13), it can be seen that $\hat{\beta}_3 > \hat{\beta}_2 > \hat{\beta}_1$, then we conclude that factor 3 (LQ raw materials) causes the highest backorder level, followed by factor 2 (MQ raw material) and then factor 1 (HQ raw materials). Furthermore, factors 1, 2 and 3 have inverse effects because $\hat{\beta}_{12}$, $\hat{\beta}_{13}$ and $\hat{\beta}_{23}$ are negative, then when factors (1 and 2), (1 and 3) or (2 and 3) are used, the resulting backorder has a lower average than expected by averaging the backorder level by the pure factor 1, 2 or 3. It means using only HQ, MQ or LQ raw materials causes the backorder level to increase.

Figure 5 plots the contours of backorder level using Design-Expert 8. In the figure, the shading comes out from blue to orange according to six defined ranges. The white point in the figure represents the backorder value of the ideal supply and red points are the backorder value of other scenarios. For the cost and inventory level we use the above definitions as well. Moreover Appendix reports the experimental matrix and the response of the mixture design of KPIs.

In Figure 5, it can be seen that x_1 (HQ) has the greatest effect on minimizing the backorder. On the plot, start from the pure factor x_1 (vertex (A: HQ)) and go to the centre of the edge towards vertex (C: LQ). The backorder level in the mixture made by 57% HQ and 43% LQ is in the cobalt blue range; however, when LQ receives to the pure factor in vertex (C: LQ) with orange range, the backorder level becomes a threat to the benefit of the company. On the opposite side, start from the pure factor x_1 (vertex (A: HQ)) and go to the centre of the edge towards vertex (B: MQ). The backorder level in the mixture made by 57% HQ and 43% MQ is in the cobalt blue range like the LQ class. In this situation, when MQ percentage moves to the pure factor in vertex (B: MQ), the backorder level becomes a problem, although it is in the green range. This shows the serious effect of lack of HQ class.

Table 2. The scenarios in defined ranges in the backorder level.

	Results of backorder level					
Experiments	Extremely Desirable	Desirable	Acceptable	Undesirable	Unacceptable	Extremely Unacceptable
Ideal	SI	–	–	–	–	–
Experiments 1	S1, S2, S3, S4, S5, S6	S7	S8	S9, S10, S11, S12	–	–
Experiments 2	S13, S14, S15, S16, S17	S18, S19	S20	S21, S22, S23, S24	–	–
Experiments 3	S25, S26, S27	S28, S29, S30, S31, S32			–	–
Experiments 4	S33, S34, S35	–	–	–	–	–
Experiments 5	–	–	S36	–	S37	S38

As the minimum backorder quantity is needed, the scenarios are classified in considered ranges based on their estimated values (see Table 2). Twenty-six scenarios are distinguished with better performances than *Acceptable* range in terms of the backorder quantity. Among these scenarios, 18 scenarios lie in *Extremely Desirable* range. In Experiments 1 and 2, the scenarios with less than 30% decrease in HQ class belong to *Extremely Desirable* or *Desirable* ranges. It means that the decrease rate till 25% of HQ classes keeps the backorder quantity in the desirable range. In Experiments 3, the scenarios with less than 30% decrease in MQ class belong to *Extremely Desirable* or *Desirable* ranges. Then, the decrease rate by 25% of MQ classes keeps the backorder quantity in the desirable range. All scenarios in Experiments 4 are put in *Extremely Desirable* range; this means that decrease by 12% in LQ classes does not have any effect in the backorder quantity. None of the scenarios belonging to Experiments 5 lie in the desirable range, meaning that using the pure factors results in growth of the backorder size.

A 3D surface for the backorder function with 95% confidence interval is presented in Figure 6. We design 38 scenarios for experiment; however, the 3D surface in Figure 6 gives

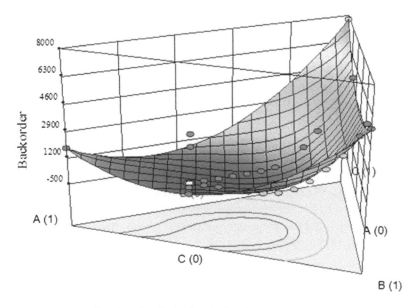

Figure 6. Three-dimensional surface for the backorder level.

us a general estimation about the behaviour of the other new scenarios which are not designed. The dots in the surface show 100% fitted scenarios. The dots up or below the model surface are fitted by an error. The triangle that shows the contour plot of the back-order level was explained in Figure 5.

Since dynamic market is a more important characteristic of the case under study and minimum service level variability (minimum backorder size) has a crucial role in this case, the safety threshold for this KPI should be defined with more sensitivity. Here, the optimal setting is considered where HQ classes have less than 30% reduction related to the ideal supply (only blue region). In this situation the backorder size has to be in *Extremely Desirable* range so that the company can deal with supply changes in this KPI.

According to the above discussion, the safety threshold of supply variations for this KPI is observed among scenarios S1, S2, S3, S4, S5, S6, S13, S14, S15, S16, S17, S25, S26, S27, S33, S34 and S35 (a total of 17 scenarios). Other remaining scenarios (21 scenarios) cross from the safety threshold and become threatening to the benefit of company. Indeed, planners should avoid accepting the mixes of these scenarios. These analyses also show that if the planners do not have any control over the supply variations and are faced with these 21 scenarios, the backorder level becomes a problem for the company.

4.3.2. Cost value

When several qualities of raw materials are applied, it is expected that the lower-quality products would have lower total costs. However, this expectation is not always met in the context of the remanufacturing industry. The operation costs of defective raw materials (including wastage costs) depend on the amount of imperfection. Although lower-quality lumber has lower material cost, it often needs more input materials and processing efforts to make the same number of usable parts in comparison with higher quality lumber. Thus, lower-quality lumber is less expensive to purchase, but sometimes it incurs higher costs when processing. Purchasing low-quality raw materials is thus not necessarily cost-effective for the mill, whereas the production costs of some of this cheap lumber is more than or the same as production costs of expensive lumber. On the other hand, buying high-quality raw materials (which is a common practice in such mills) is followed by high procurement costs. In fact, how to balance the required quality in a cost-effective manner to fulfil customer demand is very essential. For this purpose, this subsection seeks scenarios with the minimum costs.

From the company's point of view, scenarios with lower costs are thus admitted. Therefore, we studied the possible models (linear, quadratic, special cubic, cubic, special quartic and quartic) to show the response surface of the cost using Design-Expert software 8.0.7.1 (Stat-Ease 2011). According to ANOVA results, a special cubic model with a large predicted R-squared (0.7812) and the smallest predicted residual sum of squares (PRESS) statistic was chosen to fit the response surface of the cost (Equation (14)). The parameters of this equation are defined the same as Equation (11), with this difference that $\hat{\beta}_{123}$ is the coefficient of the interaction terms 1, 2 and 3.

$$\hat{y}' = \hat{\beta}_1 x_1 + \hat{\beta}_2 x_2 + \hat{\beta}_3 x_3 + \hat{\beta}_{12} x_1 x_2 + \hat{\beta}_{13} x_1 x_3 + \hat{\beta}_{23} x_2 x_3 + \hat{\beta}_{123} x_1 x_2 x_3 \qquad (14)$$

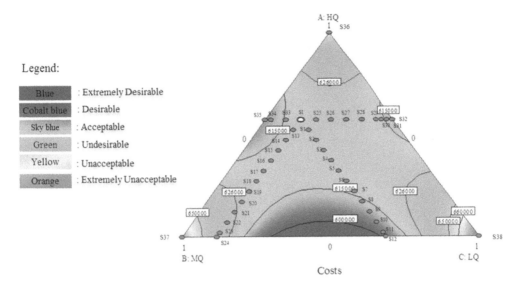

Figure 7. Contours of the cost value.

By doing the experiments, the coefficients of the equation are estimated for the cost response (Equation (15)).

$$
\text{Estimated cost function} = 656895\, x_1 + 672359\, x_2 + 683166\, x_3 - 221282 x_1\, x_2 \\
- 226107\, x_1\, x_3 - 359918 x_2\, x_3 + 103033 x_1\, x_2\, x_3
\tag{15}
$$

As P-value for coefficients' two-term and three-term interactions is less than 0.0001 in this model, these coefficients are significant and cannot be removed from the model. Note that because $\hat{\beta}_3 > \hat{\beta}_2 > \hat{\beta}_1$, we conclude that factor 3 (LQ raw materials) causes the highest costs (including production costs and inventory holding costs), followed by factor 2 (MQ raw material) and then factor 1 (HQ raw materials). These results confirm that lower-quality lumber is less expensive to purchase, but creates higher costs when processing.

Although the difference of these coefficients is small, they confirm that inferior quality lumber is less expensive to purchase, but result in higher costs when processing, particularly while using the entire production capacity. Furthermore, factors 1, 2 and 3 have antagonistic effects because $\hat{\beta}_{12}$, $\hat{\beta}_{13}$ and $\hat{\beta}_{23}$ are negative, then, when factors (1 and 2), (1 and 3) or (2 and 3) are used, the resulting costs have a lower average than expected by

Table 3. The scenarios in defined ranges in the cost value.

			Results of cost value			
Experiments	Extremely Desirable	Desirable	Acceptable	Undesirable	Unacceptable	Extremely Unacceptable
Ideal	–	–	SI	–	–	–
Experiments 1	S11, S12	S7, S8, S9, S10	S1, S2, S3, S4, S5, S6	–	–	–
Experiments 2	–	–	S13, S14, S15, S16, S17, S18, S19	S20, S21, S22, S23, S24	–	–
Experiments 3	–	S31, S32	S25, S26, S27, S28, S29, S30	–	–	–
Experiments 4	–	S34, S35	S33	–	–	–
Experiments 5	–	–	–	–	S36	S37, S38

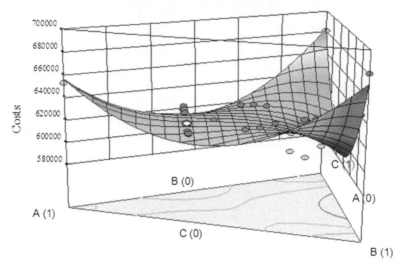

Figure 8. Three-dimensional surface for the cost value.

averaging the costs value by the pure factor 1, 2 or 3. A three-term interaction is positive so that it has an additive effect in costs.

Figure 7 plots the contours of cost value. Vertex (A: HQ) is in the green range and vertex (B: MQ) and vertex (C: LQ) with orange colour have higher cost. On the plot, start from the pure factor x_1 (vertex (A: HQ)) and go towards the centre of the triangle. The cost value in the mixtures made by 57% HQ is in the sky blue range. The figure demonstrates that mixes of quality supply have good results in this KPI; whereas the pure supply HQ, MQ and LQ raw material classes are not acceptable. However, as we mentioned in the introduction, the company often tends to purchase higher quality lumber to enable it to cover all quality types of orders. These results confirm that this industrial practice is expensive and there have been other mixes of supply quality with better costs.

Table 3 shows the scenarios classification according to the model estimated value. Ten scenarios are distinguished with better performances than *Acceptable* range in terms of cost.

A 3D surface for the cost function is presented in Figure 8 in which dots on the surface show 100% fitted scenarios and dots above or below the surface are fitted by an error. Under this model the optimal setting is where HQ classes decrease by 35% around the ideal supply and are replaced with MQ or LQ class. But as purchasing low-quality lumber is not necessarily cost-effective for the mill, the costs variations do not make a big difference (in contrast to the backorder size). The company's point of view is to minimize the cost; therefore the safety threshold of this KPI is defined for *Extremely Desirable, Desirable* and *Acceptable* ranges, meaning that the scenarios in other ranges are not admitted. It is interesting to note that the acceptable range belongs to the scenarios with fewer backorder quantities. In the contrast, scenarios with minimum costs (such as S12, S11 and S10) are mixes of MQ and LQ materials that cannot meet the customer demand but they have high level of inventories as we show in the next section. The reason is that without HQ materials, demand fulfilment is not

86

completely possible and an increase in percentage of MQ and LQ materials only causes high level of inventory.

The safety threshold of supply variations for the cost as the second KPI is observed among scenarios S11, S12, S7, S8, S9, S10, S31, S32, S34, S35, S1, S2, S3, S4, S5, S6, S13, S14, S15, S16, S17, S18, S19, S25, S26, S27, S28, S29, S30 and S33 (a total of 30 scenarios). Other remaining scenarios (eight scenarios) cross from the safety threshold and become threatening to the benefit of company for this KPI. Planners should avoid accepting the mixes in these scenarios. If planners do not know the impact of supply variations, these scenarios lead them to a problem in terms of costs.

4.3.3. Inventory level

Finding scenarios with the lower inventory level is the objective of this subsection. To fit a response surface for the inventory, the ANOVA suggested a quartic model with a large predicted R-squared (0.8395) (Equation (16)).

$$\hat{y}'' = \hat{\beta}_1 x_1 + \hat{\beta}_2 x_2 + \hat{\beta}_3 x_3 + \hat{\beta}_{12} x_1 x_2 + \hat{\beta}_{13} x_1 x_3 + \hat{\beta}_{23} x_2 x_3 + \hat{\beta}_{1122} x_1 x_2 (x_1 - x_2)$$
$$+ \hat{\beta}_{1133} x_1 x_3 (x_1 - x_3) + \hat{\beta}_{111222} x_1 x_2 (x_1 - x_2)^2 + \hat{\beta}_{222333} x_2 x_3 (x_2 - x_3)^2 \quad (16)$$

By doing the experiments, the coefficients of the equation are estimated according to Equation (17).

$$\text{Estimated inventory function} = 172164 x_1 + 176896 x_2 + 178043 x_3 - 938.06 x_1 x_2 + 6270.67 x_1 x_3$$
$$- 34595.60 x_2 x_3 - 37755.36 x_1 x_2 (x_1 - x_2) + 46716.92 x_1 x_3 (x_1 - x_3)$$
$$- 90121.28 x_1 x_2 (x_1 - x_2)^2 + 61708.02 x_2 x_3 (x_2 - x_3)^2$$
$$(17)$$

Note that because $\hat{\beta}_3 > \hat{\beta}_2 > \hat{\beta}_1$, we conclude that factor 3 (LQ raw materials) causes the highest backorder level, followed by factor 2 (MQ raw material) and then factor 1 (HQ raw materials). Furthermore, factors 1, 2 and 3 have antagonistic effects because $\hat{\beta}_{12}$ and $\hat{\beta}_{23}$ are negative, then when factors (1 and 2) or (2 and 3) are used, the resulting

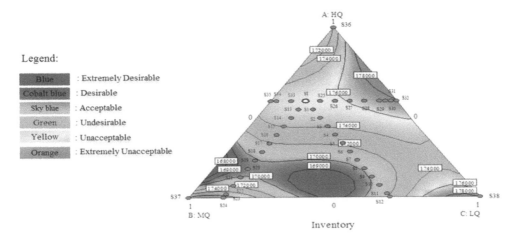

Figure 9. Contours of inventory level for all experiments.

Table 4. The scenarios results of the inventory level.

			Results of inventory level			
Experiments	Extremely Desirable	Desirable	Acceptable	Undesirable	Unacceptable	Extremely Unacceptable
Ideal	–	–	–	–	SI	–
Experiments 1	–	–	S6, S7, S8, S9, S10, S11	S3, S4, S5, S12	S1, S2	–
Experiments 2	–	S19, S20, S21	S17, S18, S22	S15, S16	S13, S14	S23, S24
Experiments 3	–	–	–	–	S25, S26	S27, S28, S29, S30, S31, S32
Experiments 4	–	–	–	–	S33, S34, S35	
Experiments 5	–	–	–	S36	–	S37, S38

inventory has a lower average than expected by averaging the inventory level by the pure factor 1, 2 or 3. The factors (1 and 3) in $\hat{\beta}_{13}$ and $\hat{\beta}_{1133}$ and factors (1 and 2) in $\hat{\beta}_{222333}$ are positive, then they have additive effects in the inventory level.

Figure 9 plots the contours of the inventory level. Vertex (C: LQ) and vertex (B: MQ) are in the orange range and vertex (A: HQ) lies in the green range. While the percentage of HQ class of raw materials is gradually decreased (from the colour yellow in the ideal point going down towards the edge HQ = 0%), the inventory of end-products is gradually decreased. While HQ percentage is decreased, the HQ production rate is consequently decreased and often its end-products are allocated to the customer orders. On the other hand, production based on the MQ and LQ is going to start. Therefore at the point of 30% HQ class, the end-products have the minimum inventory (blue colour). After this point the production capacity is often applied to produce wrong products with LQ or MQ class so the inventory level goes up again (in the edge of HQ = 0%). On the plot, start from the pure factor x_1 (vertex (A: HQ)) and go towards actor x_3 (vertex (C: LQ)). The inventory level in the mixture made by 57% HQ and 43% LQ is in the orange range because of producing the wrong products by LQs.

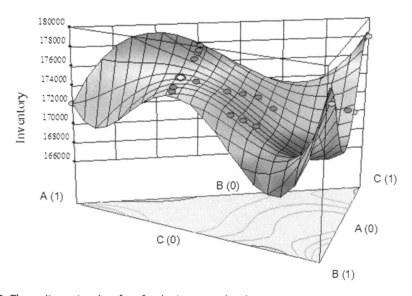

Figure 10. Three-dimensional surface for the inventory level.

Table 4 shows the results of the inventory based on the model estimated values. Three scenarios are distinguished with better performances than *Acceptable* range in terms of the inventory quantity.

A 3D surface for the inventory function is presented in Figure 10. Dots on the surface show 100% fitted scenarios and dots over or below the surface are fitted by an error.

In co-production systems, the task of inventory control is very complicated. Even the ideal supply in this case (which is obtained by the objective function of backorder minimization) is in *Unacceptable* range in terms of the inventory. As the inventory holding cost is relatively low in this industry and high inventory quantities of (undesired) co-products are always observed, the sensitivity of this KPI is less than other KPIs in this case. Therefore, the safety threshold of inventory is defined where the production of wrong products are prevented. As a result, the inventory level should be at least in *Unacceptable* range. In Figure 10, the orange range will be the reject region.

In terms of inventory level, the safety threshold of supply variations is observed among scenarios S19, S20, S21, S6, S7, S8, S9, S10, S11, S17, S18, S22, S3, S4, S5, S12, S15, S16, S36, S1, S2, S13, S14, S25, S26, S33, S34 and S35 (a total of 28 scenarios). Other remaining scenarios (10 scenarios) cross from the safety threshold and become threatening to the benefit of company in this KPI. Planners should avoid accepting the mixes in these scenarios. If planners do not know the impact of supply variations, these scenarios lead them to the problem in this KPI.

4.3.4. Graphical optimization for all KPIs

So far the admissible regions for each individual KPI have been determined. In this section the goal is to seek a region of admissible responses for all KPIs. To achieve this goal, we specify the highest acceptable upper limit for each KPI. The graphical optimization, referred to as an overlay graph, is comprised of the contour plots from each response laid

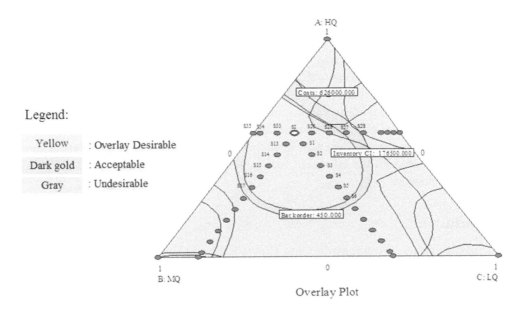

Figure 11. Overlay plot of the backorder level, cost value and inventory level.

Table 5. The comparison between real observations and estimated values in optimum region.

Experiments	Optimum regions (yellow and dark golden)	
	Real observation	Estimated value
Ideal	SI	SI
Experiments 1	S1, S2, S3	S1, S2, S3, S4, S5, S6
Experiments 2	S13, S14, S15, S16, S17	S13, S14, S15, S16, S17
Experiments 3	S25, S26, S27	S25, S26, S27
Experiments 3	S33, S34, S35	S33, S34, S35

on top of each other. A 95% confidence interval is set to obtain the admissible region for every KPI. We set the criteria for each individual response. The backorder level safety threshold is in *Desirable* range (450 units), the cost value safety threshold is in the *Undesirable* range (till 626,000 $) and the inventory level safety threshold is in *Extremely Unacceptable* range (till 176,500 units).

Figure 11 shows the overlay plot of the three response surfaces. The optimal region (yellow colour) shows where the entire ranges of all intervals meet the specified criteria. The dark gold region corresponds to where the estimated points meet the criteria requirements, but not for all parts of the criteria. The optimum region includes 16 scenarios with the desirable overlay and three scenarios with results close to this region (dark gold coloured region) which can be conditionally admissible.

As a summary, the safety threshold of supply variations for all KPIs is observed among scenarios S1, S2, S3, S4, S5, S6, S13, S14, S15, S16, S17, S25, S26, S33, S34 and S35. These 16 scenarios lie in the optimum region for all KPIs and they are considered as overall safety threshold for the supply variation in the case under study. Scenarios S7, S18 and S27 also have the results close to the optimal and can be considered as the secondary acceptable scenarios. Other remaining scenarios (19 scenarios) cross from the safety threshold and become threatening to the benefit of company with respect of all KPIs.

These analyses clearly show that without awareness of the impacts of supply variation on production planning, decision-makers may choose mixes of supply quality that not only do not have benefit for the company, but also they are not able to meet customer demand and only increase the level of inventory. These scenario analyses give insights to decision-makers about how to handle the supply variations in their environment.

Although selecting the presented surface models was based on the smallest predicted residual sum of squares, these models like other estimated models include the errors. In Table 5, the results of real observations and estimated values of surface models are compared in order to show errors of the estimated models. The estimated values are fitted by real observations in most of the cases. Scenarios 4, 5 and 6 in terms of the costs have significant differences with the estimated cost values; therefore, they lie beyond the considered safety threshold.

5. Conclusion

This article applies an optimization and simulation tool to analyze the impact of supply uncontrollable in a real-scale wood remanufacturing mill in Eastern Canada. Since the mill is a remanufacturing business in which suppliers are all owned by one holding company, planners are understandably not able to predict quality of raw materials, although they may control quantities. In such a mill, it is not easy to say how much perturbation in supply becomes just a problem or a threat for the benefit of company. The present contribution using an experimental platform helps managers make decisions about how to deal with uncontrollable supply, under the complex characteristics of the wood remanufacturing industry. By comparing different scenarios for perturbed supply through simulation/ optimization experiments, the proposed methodology allows managers to identify the desirable region of mixed lumber quality based on KPIs. The obtained results determine an optimum region including a series of good enough scenarios based on the company's point of view to make managers aware of the impacts of supply uncontrollability on the production planning processes.

Although many papers have focused on production planning optimization in the forest supply chain during the past few years, there is no existing methodology available to solve the wood remanufacturing industrial problem formulated in this paper. Indeed the complex planning task (e.g. divergent co-production, alternative processes, a make-to-order philosophy, short order cycle times, and uncontrollable supply) which comes from inherent characteristics of the wood remanufacturing industry is the contribution if this paper. Based on this research, the synergy between the primary processing industry and the remanufacturing industry is a good opportunity to ensure that they work together towards a common objective. Thus creating appropriate links between the primary processing industry and the secondary industry could offer opportunities for improvements. Therefore, it is interesting to provide collaboration mechanisms between such interdependent companies.

Disclosure statement

No potential conflict of interest was reported by the authors.

Funding

This work was supported by the Natural Sciences and Engineering Research Council of Canada (NSERC) Strategic network on Value Chain Optimization (VCO) [grant number 387200-09].

References

Boyabatli O, Kleindorfer PR, Koontz SR. 2011. Integrating long-term and short-term contracting in beef supply chains. Manag Sci. 57:1771–1787.

Buck RA, Buehlmann U, Thomas RE. 2010. Romi-3.1 least-cost lumber grade mix solver using open source statistical software. Forest Prod J. 60:432–439.

Buehlmann U, Thomas RE. 2001. Lumber yield optimization software validation and performance review. Rob Comput Integr Manuf. 17:27–32.

Buehlmann U, Thomas RE, Zuo XQ. 2011. Cost minimization through optimized raw material quality composition. Rob Comput Integr Manuf. 27:746–754.

Buehlmann U, Zuo XQ, Thomas RE. 2010. Second-order polynomial model to solve the least-cost lumber grade mix problem. Forest Prod J. 60:69–77.

Cornell JA. 2011. A primer on experiments with mixtures. New York: John Wlley & Sons.

D'Amours S, Rönnqvist M, Weintraub A. 2008. Using operational research for supply chain planning in the forest products industry. INFOR Inform Syst Oper Res. 46:265–281.

Denizel M, Ferguson M, Souza GC. 2010. Multi-period remanufacturing planning with uncertain quality of inputs. IEEE Trans Eng Manage. 57:394–404.

Feng Y, D'Amours S, Beauregard R. 2010. Simulation and performance evaluation of partially and fully integrated sales and operations planning. Int J Prod Res. 48:5859−5883.

Gaudreault J, Forget P, Frayret JM, Rousseau A, D'Amours S. 2010. Distributed operations planning in the lumber supply chain: models and coordination. Int J Ind Eng. 17(3): 168−189.

Harding OV, Steele PH. 1997. Rip-x: decision software to compare crosscut-first and rip-first rough mill systems. Wood Sci Technol. 31:367−381.

Hoff K, Fisher N, Miller S, Webb A. 1997. Sources of competitiveness for secondary wood products firms: a review of literature and research issues. Forest Prod J. 47:31−37.

Itami RM, Zell D, Grigel F, Gimblett HR. 2005. Generating confidence intervals for spatial simulations: determining the number of replications for spatial terminating simulations. In: Proceedings of the MODSIM 2005 − International Congress on Modelling and Simulation: Advances and Applications for Management and Decision Making. Melbourne, Australia; p. 141−148.

Kazemi Zanjani M, Nourelfath M, Ait-Kadi, D. 2011. Production planning with uncertainty in the quality of raw materials. J Opl Res Soc. 62(7):1334−1343.

Kozak RA, Maness TC, Caldecott T. 2003. Solid wood supply impediments for secondary wood producers in British Columbia. Forest Chron. 79:1107−1120.

Law AM, Kelton WD. 2007. Simulation modeling and analysis, 4th edition. Boston/London: McGraw-Hill.

Lihra T. 2007. Synergy between primary processing and remanufacturing. Conference of innovating in wood remanufacturing: status review and prospects on the Gaspé Peninsula; Quebec, Canada.

Mukhopadhyay SK, Ma H. 2009. Joint procurement and production decisions in remanufacturing under quality and demand uncertainty. Int J Prod Econ. 120:5−17.

Myers RH. 1999. Response surface methodology − current status and future directions - response. J Qual Technol. 31:73−74.

Ostermark R. 1999. Solving a nonlinear non-convex trim loss problem with a genetic hybrid algorithm. Comput Oper Res. 26:623−635.

Rafiei R, Nourelfath M, Gaudreault J, Santa-Eulalia LA, Bouchard M. 2014. A periodic re-planning approach for demand-driven wood remanufacturing industry: a real-scale application. Int J Prod Res. 52:4198−4215.

Rafiei R, Nourelfath M, Gaudreault J, Santa-Eulalia LA, Bouchard M. 2015. Dynamic safety stock in a co-production demand-driven wood remanufacturing mill: a case study. Int J Prod Econ. 165:90−99.

Rönnqvist M, Astrand E. 1998. Integrated defect detection and optimization for cross cutting of wooden boards. Eur J Oper Res. 108:490−508.

Santa-Eulalia LA, Aït-Kadi D, D'Amours S, Frayret J-M, Lemieux S. 2011. Agent-based experimental investigations about the robustness of tactical planning and control policies in a softwood lumber supply chain. Prod Plan Control. 22:782−799.

Schmincke KH. 1995. Forest industries: crucial for overall socio-economic development. International Journal of Forestry and Forest Industries. Published by the Food and Agriculture Organization of the United Nations, vol. 46.

Shi J, Zhang G, Sha J. 2011. Optimal production planning for a multi-product closed loop system with uncertain demand and return. Comput Oper Res. 38:641−650.

Wang CX. 2009. Random yield and uncertain demand in decentralised supply chains under the traditional and VMI arrangements. Int J Prod Res. 47:1955−1968.

Appendix

Table A1. Design matrix for three-factor mixture design.

Scenarios		Component 1 (HQ)	Component 2 (MQ)	Component 3 (LQ)	Response 1 (Backorder QTY)	Response 2 (Cost)	Response 3 (Inventory QTY)
Ideal	SI	0.57	0.31	0.12	43	621,047	175,988
Experiments 1	S1	0.52	0.31	0.17	44	608,990	175,521
	S2	0.47	0.31	0.22	47	621,457	174,441
	S3	0.42	0.31	0.27	58	615,030	174,381
	S4	0.37	0.31	0.32	73	632,542	174,086
	S5	0.32	0.31	0.37	98	631,743	173,675
	S6	0.27	0.31	0.42	215	629,231	172,424
	S7	0.22	0.31	0.47	579	604,601	168,663
	S8	0.17	0.31	0.52	1261	601,221	168,642
	S9	0.12	0.31	0.57	2184	612,480	170,152
	S10	0.07	0.31	0.62	3176	611,962	172,117
	S11	0.02	0.31	0.67	4597	608,394	171,903
	S12	0.00	0.31	0.69	5295	605,040	171,702
Experiments 2	S13	0.52	0.36	0.12	44	615,954	174,290
	S14	0.47	0.41	0.12	78	614,570	173,716
	S15	0.42	0.46	0.12	61	615,757	173,329
	S16	0.37	0.51	0.12	79	620,274	173,134
	S17	0.32	0.56	0.12	105	623,025	172,840
	S18	0.27	0.61	0.12	267	627,835	172,518
	S19	0.22	0.66	0.12	638	604,787	168,433
	S20	0.17	0.71	0.12	1281	600,715	168,400
	S21	0.12	0.76	0.12	2171	611,969	169,264
	S22	0.07	0.81	0.12	3136	633,723	173,542
	S23	0.02	0.86	0.12	4442	640,167	176,185
	S24	0.00	0.88	0.12	5203	641,051	175,865
Experiments 3	S25	0.57	0.26	0.17	45	618,699	175,450
	S26	0.57	0.21	0.22	46	621,351	175,717
	S27	0.57	0.16	0.27	47	623,523	175,844
	S28	0.57	0.11	0.32	54	624,371	176,817
	S29	0.57	0.06	0.37	125	624,749	177,198
	S30	0.57	0.04	0.39	405	620,549	177,372
	S31	0.57	0.02	0.41	829	615,029	177,778
	S32	0.57	0.00	0.43	1580	597,060	177,561
Experiments 4	S33	0.57	0.36	0.07	44	616,604	175,107
	S34	0.57	0.41	0.02	45	616,677	175,566
	Hq	0.57	0.43	0.00	45	618,347	175,246
Experiments 5	S36	1.00	0.00	0.00	1826	652,902	172,156
	S37	0.00	1.00	0.00	5532	680,850	176,899
	S38	0.00	0.00	1.00	7733	681,000	177,999

Development of a threat index to manage timber production on flammable forest landscapes subject to spatial harvest constraints

Juan J. Troncoso, Andrés Weintraub and David L. Martell

ABSTRACT

We develop a stand-level fire threat index and incorporate it in a mixed integer programming model that can be used to help specify strategies that will maximize the expected volume harvested from a flammable forest over a finite planning horizon, subject to adjacency constraints. The inclusion of a threat index in our objective function implicitly identifies high volume harvest blocks that are most likely to burn over the planning horizon and accelerates the harvesting of those high risk stands to reduce the likelihood that they will burn before they are harvested. We illustrate the use of our model and evaluate its performance by simulating its application to two hypothetical flammable forest landscapes that are 16,000 hectares in size. Our results indicate that our inclusion of a threat index in our objective function produces timber harvest schedules that are better than those produced using an objective function that does not include potential fire loss measures, while satisfying harvest adjacency constraints.

1. Introduction

Linear programming (LP) models have been widely used for strategic forest harvest scheduling since the early 1970s when Navon (1971) developed Timber Ram for the US Forest Service. With the exception of Balas' seminal mixed integer programming formulations of timber harvesting and road network planning problems in Romania during the period 1959–1964 (described by Rand 2006), most early timber harvest scheduling models were aspatial models. Individual stands were aggregated into relatively homogeneous (with respect to, for example, age, cover type and site quality) strata that served as the basic unit of analysis. The optimal solutions to such models specified how much of each stratum to harvest during each period, but not when specific stands within those strata were to be cut. The solutions to aspatial LP models were provided to experienced forest managers who were expected to subjectively implement them on the landscape by 'allocating the cut' – by determining which stands within each stratum would be harvested during each period.

Concern about the aesthetics and ecological impact of large clear-cuts created a demand for decision support systems that could be used to help generate and evaluate spatially explicit forest harvest schedules, one of the simplest and earliest versions of which is commonly referred to as the *adjacency problem*. Early versions of the adjacency problem called for partitioning the forest into cut blocks, the size of each of which was less than or equal to some specified maximum allowable harvest block size. The decision-making problem was to determine when each cut block was to be harvested subject to the constraint that no two adjacent cut blocks could be harvested during some specified green-up period. Most adjacency problems were formulated as integer or mixed integer programming problems that were solved using both exact (e.g. Barahona et al. 1992; Snyder & ReVelle 1996; Goycoolea et al. 2005) and heuristic methods (e.g. Lockwood & Moore 1993).

Weintraub and Murray (2006) is a comprehensive review of adjacency and other spatially explicit mathematical programming forest harvest scheduling models that had been published up until 2005. Murray (1999) classified most such spatial harvest planning models as being either what he described as unit restriction models (URMs) or area restriction models (ARMs). URMs are designed to schedule the harvest of cut blocks (the size of all of which are less than or equal to the maximum allowable cut block size but large enough so that no two adjacent blocks can be harvested together without violating the adjacency constraints). Cut blocks are formed from basic units, typically homogeneous forest stands. Often these blocks are formed manually by a forest planner using a geographic information system. ARMs include in the decision model, both the creation of the cutting blocks and scheduling of the harvesting of them to satisfy cut block size and adjacency constraints (see Carvajal et al. 2013). The latter approach creates models that produce better solutions but are computationally far more difficult to solve. In this paper we consider the first (URM) approach, where the model receives the cutting block descriptions as input data.

Many authors have investigated the development and use of fuel management models that can be used to help decide when and where to treat flammable fuels to reduce the flammability of forest landscapes (e.g. Omi et al. 1981; Wei et al. 2008; Kim et al. 2009; Rytwinski & Crowe 2010; Yoshimoto et al. 2010; Konoshima et al. 2010; Minas et al. 2014) and some (e.g. Van Wagner 1983; Reed & Errico 1986; Gassmann 1989; Boychuk & Martell 1996; Armstrong 1999, 2004; Savage et al. 2010; Chung 2015; Ferreira et al. 2015) have developed and used aspatial models to investigate the impact of fire on timber supply. Others (e.g. Bettinger 2009; Acuña et al. 2010) have included uncertain fire losses in spatially explicit forest harvest scheduling models but we are not aware of any efforts to develop and use mathematical programming models to develop spatially explicit harvest scheduling models that satisfy adjacency constraints on flammable landscapes on which fire can create patches which, when coupled with clear-cuts, violate adjacency constraints.

1.1. The decision-making problem

In this paper we focus on the problem of determining how best to manage the harvesting of a flammable forest over a finite planning horizon subject to both adjacency and harvest flow constraints. Deterministic spatial planning models are of questionable value to forest managers that must satisfy spatial constraints when they plan the harvesting of flammable forests because it is possible that some of the stands they plan to harvest and some of the stands that are adjacent to the stands they plan to harvest may burn before they are harvested. We address the problem of determining when to harvest stands to maximize the volume harvested and satisfy both harvest flow and adjacency constraints. There are, to our knowledge, no methods for deriving optimal solutions to such problems. Our approach is based upon on the principle of FireSmart forest management articulated in Hirsch et al. (2001) and investigated in Acuña et al. (2010). Following Hirsch et al. (2001), we assume that fire cannot ignite in or spread through any stand that has been harvested.

1.2. Overview

We began by formulating a simple hypothetical spatial harvest scheduling problem that is representative of the types of decision-making problems that forest managers must resolve when they develop spatially explicit harvest schedules for flammable forests. We model the management of our forests over a 50-year planning horizon that is partitioned into five 10-year periods and address spatial considerations by imposing URM type (Murray 1999) adjacency constraints on harvesting activities. We do not include the regeneration of cut blocks in our model because we assume the forest may be used for some purpose other than timber production (e.g. agriculture) beyond the end of the planning horizon, so we are interested in the state of our forest during and at the end of that planning horizon but not how it is used or managed thereafter.

We evaluated our model by simulating its implementation and determining how well it would perform where it applied to a hypothetical 16,000 hectare flammable forest landscape that comprises 400 square stands, each of which is 40 hectare in size, embedded in a uniform grid. We consider two different landscape configurations. Our first landscape (Figure 1) comprises three distinct groups or zones of cutting blocks that differ by age, merchantable volume and fire ignition probability, but each zone is relatively homogeneous with respect to the ignition probability of its cut blocks. The ignition probabilities of all the cells in the first (low hazard) zone are uniformly distributed over the range 0.00–0.02. Those in the second (medium hazard) zone have ignition probabilities that are uniformly distributed over the range 0.02–0.04 and the third (high hazard) zone contains cells for which the ignition probability is uniformly distributed over the range of 0.04–0.06 The second landscape, which we refer to as the heterogeneous landscape which is depicted in Figure 2, also contains three (low, medium and high hazard) zones, but those zones are fragmented into many homogeneous compartments that range in size from 1 to as many as 37 cut blocks. The growth and yield functions used in this case study were for a *Nothofagus pumilio* forest (native forest) owned by a Chilean forest company described in Cruz et al. (2007).

$$\text{Vol(m}^3/\text{tree}) = -1.41 \ 10^{-9} \text{age}^4 + 5.696 10^{-7} \text{age}^3 + 3.789 10^{-5} \text{age}^2 + 0.0011 \text{age} - 0.0085$$

where the age is expressed in years. To determine the volume per hectare, we considered a standard density of 500 trees per hectare.

1	2	3	4	5	6	7	8	9	10	11	12	13	14	15	16	17	18	19	20
21	22	23	24	25	26	27	28	29	30	31	32	33	34	35	36	37	38	39	40
41	42	43	44	45	46	47	48	49	50	51	52	53	54	55	56	57	58	59	60
61	62	63	64	65	66	67	68	69	70	71	72	73	74	75	76	77	78	79	80
81	82	83	84	85	86	87	88	89	90	91	92	93	94	95	96	97	98	99	100
101	102	103	104	105	106	107	108	109	110	111	112	113	114	115	116	117	118	119	120
121	122	123	124	125	126	127	128	129	130	131	132	133	134	135	136	137	138	139	140
141	142	143	144	145	146	147	148	149	150	151	152	153	154	155	156	157	158	159	160
161	162	163	164	165	166	167	168	169	170	171	172	173	174	175	176	177	178	179	180
181	182	183	184	185	186	187	188	189	190	191	192	193	194	195	196	197	198	199	200
201	202	203	204	205	206	207	208	209	210	211	212	213	214	215	216	217	218	219	220
221	222	223	224	225	226	227	228	229	230	231	232	233	234	235	236	237	238	239	240
241	242	243	244	245	246	247	248	249	250	251	252	253	254	255	256	257	258	259	260
261	262	263	264	265	266	267	268	269	270	271	272	273	274	275	276	277	278	279	280
281	282	283	284	285	286	287	288	289	290	291	292	293	294	295	296	297	298	299	300
301	302	303	304	305	306	307	308	309	310	311	312	313	314	315	316	317	318	319	320
321	322	323	324	325	326	327	328	329	330	331	332	333	334	335	336	337	338	339	340
341	342	343	344	345	346	347	348	349	350	351	352	353	354	355	356	357	358	359	360
361	362	363	364	365	366	367	368	369	370	371	372	373	374	375	376	377	378	379	380
381	382	383	384	385	386	387	388	389	390	391	392	393	394	395	396	397	398	399	400

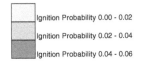

Ignition Probability 0.00 - 0.02

Ignition Probability 0.02 - 0.04

Ignition Probability 0.04 - 0.06

Figure 1. Forest in which low, moderate and high hazard cut blocks are clustered into three relatively homogeneous zones.

The age class structure of our hypothetical forest is presented in Table 1. The ignition probabilities were uniformly distributed over the interval of 0.01%–6.0%.

The forest manager must decide which blocks to harvest during each period to satisfy the demand for timber subject to adjacency constraints given uncertain future fire losses. We assume he or she will begin by considering the state of the forest at the beginning of the 50-year planning horizon, determining which stands to harvest during each of the five 10-year periods and implementing the first period solution at the start of period 1. Given the uncertain fire loss, we assume the forest manager will review the state of the forest at the end of the first period and re-plan for the remaining four periods at the start of period 2. We assume such re-assessment and re-planning will take place at the start of periods 3, 4 and 5 as well. We assume fires can ignite and spread into cells immediately adjacent to the cell in which they ignite but no further than the three adjacent cells downwind of the cell in which the fire is first ignited, and that the fire burns all of the cells in which they are ignited or into which they spread. We also assume that harvesting always occurs before any fires that do occur in any period, so any blocks that are scheduled to be harvested during a specific period will be harvested before they can burn.

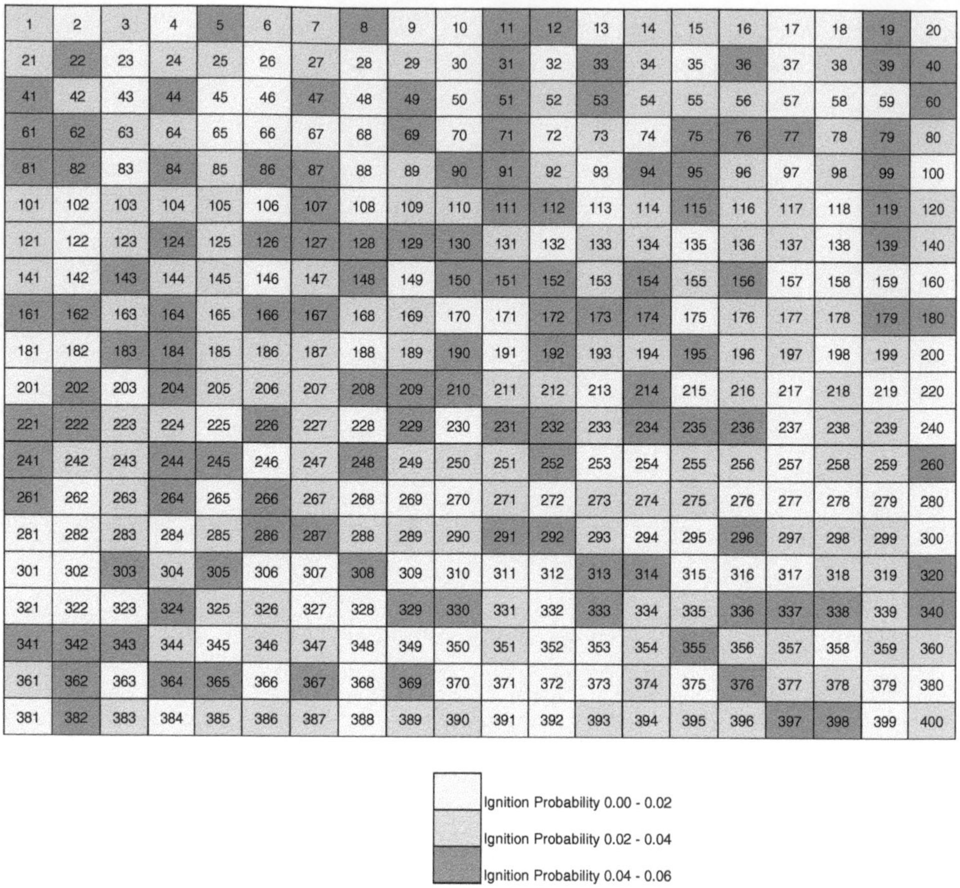

1	2	3	4	5	6	7	8	9	10	11	12	13	14	15	16	17	18	19	20
21	22	23	24	25	26	27	28	29	30	31	32	33	34	35	36	37	38	39	40
41	42	43	44	45	46	47	48	49	50	51	52	53	54	55	56	57	58	59	60
61	62	63	64	65	66	67	68	69	70	71	72	73	74	75	76	77	78	79	80
81	82	83	84	85	86	87	88	89	90	91	92	93	94	95	96	97	98	99	100
101	102	103	104	105	106	107	108	109	110	111	112	113	114	115	116	117	118	119	120
121	122	123	124	125	126	127	128	129	130	131	132	133	134	135	136	137	138	139	140
141	142	143	144	145	146	147	148	149	150	151	152	153	154	155	156	157	158	159	160
161	162	163	164	165	166	167	168	169	170	171	172	173	174	175	176	177	178	179	180
181	182	183	184	185	186	187	188	189	190	191	192	193	194	195	196	197	198	199	200
201	202	203	204	205	206	207	208	209	210	211	212	213	214	215	216	217	218	219	220
221	222	223	224	225	226	227	228	229	230	231	232	233	234	235	236	237	238	239	240
241	242	243	244	245	246	247	248	249	250	251	252	253	254	255	256	257	258	259	260
261	262	263	264	265	266	267	268	269	270	271	272	273	274	275	276	277	278	279	280
281	282	283	284	285	286	287	288	289	290	291	292	293	294	295	296	297	298	299	300
301	302	303	304	305	306	307	308	309	310	311	312	313	314	315	316	317	318	319	320
321	322	323	324	325	326	327	328	329	330	331	332	333	334	335	336	337	338	339	340
341	342	343	344	345	346	347	348	349	350	351	352	353	354	355	356	357	358	359	360
361	362	363	364	365	366	367	368	369	370	371	372	373	374	375	376	377	378	379	380
381	382	383	384	385	386	387	388	389	390	391	392	393	394	395	396	397	398	399	400

Ignition Probability 0.00 - 0.02

Ignition Probability 0.02 - 0.04

Ignition Probability 0.04 - 0.06

Figure 2. Forest that is heterogeneous with respect to the distribution of low, moderate and high hazard harvest blocks.

Table 1. Age class structure of the hypothetical forest.

Age class (years)	Mean (years)	Frequency (number of blocks)
0−10	5	0
11−20	15	0
21−30	25	0
31−40	35	41
41−50	45	77
51−60	55	96
61−70	65	120
71−80	75	66

2. Methods

2.1. Description of our approach

Our model constitutes a heuristic solution to a very complex stochastic spatial forest harvest planning problem which, as far as we know, is intractable. We cannot therefore, evaluate the difference between the solutions generated by our heuristic approach and an exact optimum. We evaluate our methods by embedding both our model and a simpler model that does not account for uncertain fire losses in a simulated forest management planning environment comprising a flammable forest that is to be managed over a 50-year planning horizon. We simulate the management of that hypothetical forest using both our model and the simpler model and compare the results they produce, to assess the extent to which the use of our model might improve the management of such forests.

The simulated management of the forests proceeds as follows. Given the uncertain fire loss, at the end of each period the forest manager will determine what was harvested and burned during the previous period, update the state of the forest and re-plan for the rest of the planning horizon. The first phase, which we refer to as the *scheduling phase*, consists of using a standard mixed integer programming spatial harvest scheduling model which includes adjacency and harvest flow constraints, that is used to determine which blocks will be scheduled for harvest during each of the next T 10-year periods. Note again, that during each period, we assume harvesting is carried out before the possible burning of cutting blocks.

The remaining *re-scheduling phases* take place within simulated re-planning scenarios. After the period 1 harvest has been decided upon and implemented, simulated fires ignite in some cells and spread to some other cells on the landscape. The characteristics of fire spread are described in Section 2.2. At the start of period 2, the blocks that were burned and harvested during period 1 are identified and the modified landscape is used as input to the harvest scheduling model to re-schedule the harvest for periods 2, 3, ... T, now a four-period planning horizon rather than the original five-period planning horizon. The blocks that are scheduled to be harvested during period 2 (which are those selected by the re-planning model, and not necessarily those that were originally scheduled for harvest during period 2 in the initial harvest scheduling phase) are then harvested before they can burn and then the period 2 simulated fire ignition and spread processes burn some blocks. At the end of period 2 or the start of period 3, a similar re-planning exercise takes place with a three-period planning horizon and so on until the start of period T or 5. Figure 3 illustrates the harvest scheduling and simulated fire events. Note that neither harvested nor previously burned blocks can be ignited or burned during future periods.

2.2. Modelling fire ignition and spread

Although there are many ways to model fire spread (e.g. Bettinger 2010; Ferreira et al. 2015), we use a very simplified version of the fire ignition and spread model developed by Wei et al. (2008) depicted in Figure 4, to model fire ignition on and spread across our landscapes. Like Wei et al. (2008), we assumed that fires are driven by prevailing north-west winds but unlike Wei et al (2008), we assumed that a fire can spread no further than the three cut blocks downwind of the cell in which it ignited.

Using Wei et al.'s (2008) fire ignition and spread model, the probability that cutting block 5 will burn can be approximated using the following expression:

$$BP_5 = Q \cdot (Ip_5 + Ip_1 \cdot ps_{15} + Ip_2 \cdot ps_{25} + Ip_4 \cdot ps_{45})$$

where

Ip_i: probability that a fire ignites in cutting block i
BP_j: probability that cutting block j burns
ps_{ij}: probability that a fire that ignites in block i will spread from block i to block j
Q: spread parameter

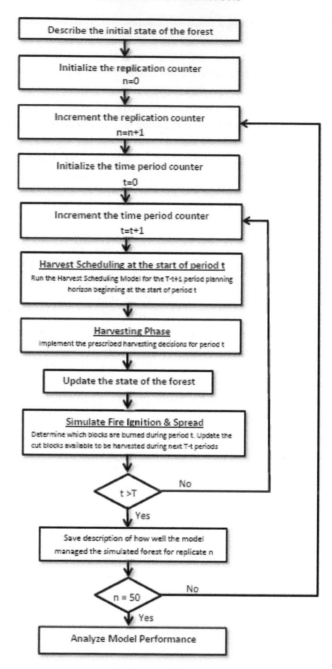

Figure 3. Schematic description of how the forest planning model is evaluated by simulating how well it performs when it is used to manage a simulated forest for 50 replications, each of which has a 50-year planning horizon that is partitioned into five 10-year periods.

The spread parameter Q was defined by Wei et al. (2008) and estimated to have a value of 0.9147. This value can vary depending on the upper bound on the fire probability in a given block. In our study, we used that same value because our burn probabilities are always less than 30% (we assumed a maximum burning probability of 10%). In case of

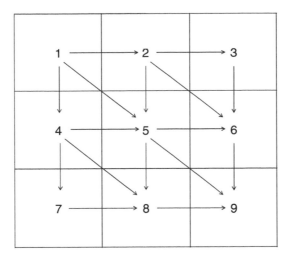

Figure 4. Representation of fire spread with a north-west wind based on Wei et al.'s (2008) fire ignition and spread model.

fire spread probabilities, they assumed that $ps_{15}/2 = ps_{25} = ps_{45}$. They also determined that ps_{15} equals 0.496 which means that if cutting block 1 is ignited and burns, this fire can spread into block 5 with a probability of 0.496. Similarly, a fire burning in block 2 or 4 would have a 24.8% chance of spreading into block 5.

We simulated fire ignition and spread as follows. Ip_i is the probability that some external agent (e.g. a person or lightning) ignites a fire in block i. Since we assume at most one fire can be ignited in block i and the ignition process in each block is independent of the ignition processes in all the other blocks on the landscape, the probability distribution of the number of fires ignited in block i is Bernoulli with an expected value of Ip_i and the expected number of fires ignited on the landscape is $\sum_1^{400} Ip_i$. We simulate fire ignition by generating 400 random numbers rn_i that are uniformly distributed over the closed interval [0,1], one for each block, and a fire ignites in cell i if $rn_i \leq Ip_i$.

We then model the spread of those fires as a simple contagion process. Consider a fire that has been ignited in block i. Following our modification of Wei et al.'s (2008) fire spread model, we assume the spread of all of the fires that have been ignited on the landscape is driven by a north-west wind and that they can only spread to the three adjacent cells that are immediately east, south-east and south of the cells in which they were ignited. To simulate fire spread we generate three random numbers rn_{ij} for cell i and since ps_{ij} is the probability that a fire that ignites in cell i can spread to cell j that is adjacent to it, a fire that is ignited in cell i will spread to cell j if $rn_{ij} \leq ps_{ij} (1 - Ip_j)$.

Note that, for simplicity, we are assuming fires that are ignited by people or lightning can spread to cells that are immediately adjacent to the cells in which they are ignited but a fire that enters a block from one of the ignition cells that is adjacent to it cannot spread any further.

2.3. Harvest scheduling model description

The model we developed to solve the harvest scheduling problem with URM spatial and green-up constraints is a typical spatial integer programming model with a modified

Table 2. Simple numerical example to illustrate the potential impact of harvesting decisions on the expected volume harvested from a flammable forest.

Block	Volume if harvested during period 1	Burn probability during period 1	Volume if not burned and harvested during period 2	Bonus 'earned' if harvested during period 1
1	100	.05	105	5
2	90	.20	100	18

objective function that maximizes the total volume plus a threat index which can be thought of as being a bonus or the expected volume harvested from cutting blocks that are harvested before they can burn. Harvesting unburned stands not only produces the volume from the stands that are harvested, it also reduces the flammability of the landscape by disrupting spread paths for future fires. We incorporated in our model, a component that will, all other things being equal, 'encourage' the harvesting of cut blocks that will have the greatest impact on reducing the subsequent volume loss due to fire. An objective function that augments the volume harvested with a bonus term will do just that.

Consider for example, a forest that comprises the two blocks described in Table 2.

If the forest was not flammable the optimal solution would be to harvest block 1 in period 1 and block 2 in period 2 to produce 200 m^3 rather than the 195 m^3 that would be produced if one harvested block 2 in period 1 and block 1 in period 2. If one accounts for the possibility that fire might burn blocks 1 and 2, the expected volume produced if one harvested block 1 in period 1 and block 2 in period 2 would be $100 + (1 - 0.20) \times 100$ or 180 m^3. The expected volume produced if one harvested block 2 in period 1 and block 1 in period 2 would be $90 + (1 - 0.05) \times 105 = 189.75$ m^3. The bonus term therefore acts as an incentive to harvesting block 2 in period 1.

Our objective function (1) below will steer harvesting towards high volume stands that are most threatened by fire. Constraints (2) ensure that any cutting block can be harvested at most once over the planning horizon. Constraints (3) are the adjacency constraints that ensure if a cut block is harvested during a particular period, none of the cut blocks that are adjacent to it can be harvested during that period. Equations (4) determine the timber production in each period. Constraints (5.a), (5.b), (6.a) and (6.b) specify the timber production requirements per period and the allowable fluctuations in harvest flow between two consecutive periods. Constraints (7.a), (7.b) and (7.c) model the impact of harvesting on the ignition probability of a stand. Equations (8.a) and (8.b) calculate the burn probabilities for each cutting block depending on the burn probabilities associated with the blocks that are adjacent to it. Finally, constraints (9.a), (9.b), (9.c), (10.a), (10.b) and (10.c) are used to determine the bonus obtained when a given cutting block is harvested which depends on the estimated burn probability for that particular block.

2.4. Mathematical structure of our harvest scheduling model

Index

Cutting blocks ... i
Planning periods ... t

Decision variables

X_{it} = binary decision variable that equals 1 if cutting block i is harvested during period t, 0 otherwise;

Auxiliary variables

$Bonus_{it}$ = continuous variable {0-1} which represents the bonus 'earned' if cutting block i is harvested during period t;

IPA_{it} = continuous variable {0-1} that determines the ignition probability of a cutting block i at the end of period t. This variable relates the ignition probability to the harvesting decisions and equates it to zero if the cutting block was harvested;

BP_{it} = burn probability calculated for cutting block i at the end of period t;

$NONHAR_i$ = continuous variable {0-1} that indicates whether or not cutting block i is harvested over the planning horizon;

$VOLTOT_t$ = total volume harvested during period t;

Parameters

vol_{it} = timber volume per hectare produced if cutting block i is harvested during period t;

$area_i$ = area available to harvest in cutting block i;

Ip_i = ignition probability for cutting block i;

ps_{ij}: fire spread probability from block i to block j;

$Initial\_BP_i$ = burn probability calculated for cutting block i at the beginning of period 1;

adj_{ij} = adjacency factor between cutting blocks i and j; equals 1 if cell i is adjacent to cell j otherwise it equals 0;

$Demandmi_t$ = lower bound for demand during period t;

$Demandma_t$ = upper bound for demand during period t;

(1) Objective function: maximize the total volume harvested plus the bonus term over the T period planning horizon:

$$\text{Max} \sum_{i=1}^{I} \sum_{t=1}^{T} vol_{it} \cdot area_i \cdot X_{it} + \sum_{i=1}^{I} \sum_{t=1}^{T} vol_{it} \cdot area_i \cdot Bonus_{it} \quad (1)$$

subject to:

(2) Harvesting constraints: a cutting block can be harvested at most once over the planning horizon.

$$\sum_{t=1}^{T} X_{it} + NONHAR_i = 1 \ \forall i \quad (2)$$

(3) URM adjacency spatial constraints: if cutting block i is harvested during period t and block j is adjacent to block i, then block j cannot be harvested during that same period. We consider both edge and corner adjacency.

$$X_{it} + adj_{ij} \cdot X_{jt} \leq 1 \ \forall i, j, t \quad (3)$$

(4) Timber production constraints: the total volume harvested during period t ($VOLTOT_t$) is the sum of all the volumes harvested from all the cutting blocks harvested during that period.

$$\sum_{i=1}^{I} vol_{it} \cdot area_i \cdot X_{it} = VOLTOT_t \ \forall t \tag{4}$$

(5) Harvest flow constraints for period 1: the total volume harvested during period 1 ($VOLTOT_t$) must be greater than or equal to the demand for timber in period 1 and less than or equal to a specified volume during that period. We assumed the demand in period 1 was 15% of the total standing volume at the beginning of the planning horizon and the upper bound on the demand in period 1 was 18% of that total standing volume (i.e. the upper bound was 20% larger than the lower bound). For the re-scheduling phases, the volume harvested in the most recently implemented period is used as a lower bound for the first re-planning period (e.g. if the re-scheduling phase starts at the beginning of period 3, the lower bound used for the volume to be harvested in that period is the volume harvested in period 2). The upper bound will be 1.2 times that lower bound.

$$VOLTOT_t \geq Demandmi_t \ t = 1 \tag{5.a}$$
$$VOLTOT_t \leq Demandma_t \ t = 1 \tag{5.b}$$

(6) Harvest flow constraints for periods 2 through 5: the volume harvested during period t must be at least 90% of the volume harvested during period t - 1 and no more than 110% of the volume harvested during period t - 1, for $t = 2, \ldots T$

$$VOLTOT_t \geq 0.9 \cdot VOLTOT_{t-1} \ \forall t \geq 2 \tag{6.a}$$
$$VOLTOT_t \leq 1.1 \cdot VOLTOT_{t-1} \ \forall t \geq 2 \tag{6.b}$$

(7) The impact of harvesting on block ignition probabilities: these three constraints model the impact of harvesting block i during period t on its ignition probability IPA_{it}. We assume that in all periods, harvesting always takes place before burning. So, if a block is not harvested during period t or earlier, its IPA_{it} will equal its ignition probability. If however, a block is harvested in period t, it cannot ignite during that period so its IPA_{it} must equal zero.

$$IPA_{it} \geq Ip_i - \sum_{e=1}^{t} X_{ie} \ \forall i, t \tag{7.a}$$

$$IPA_{it} \leq 1 - \sum_{e=1}^{t} X_{ie} \ \forall i, t \tag{7.b}$$

$$IPA_{it} \leq Ip_i \ \forall i, t \tag{7.c}$$

(8) Burn probability equations: these equations are used to determine the burn probability for each cutting block, considering ignition probabilities, and fire spread probabilities and ignition probabilities determined for each adjacent cutting block. The constraints (8.a) are used to calculate the burn probability for cutting block i at

the beginning of the planning horizon before any harvesting has taken place.

$$Initial\_BP_i = Q \cdot \left(Ip_i + \sum_{j=1}^{J} adj_{ji} \cdot Ip_j \cdot ps_{ji} \right) \quad \forall i \qquad (8.a)$$

In our study $Q = 0.9147$ similar to Wei et al.(2008)

Now, the constraints (8.b) are used to calculate the burn probability for cutting block i at the end of period t, considering the new situation for adjacent blocks that could have been harvested during the period t.

$$BP_{it} = Q \cdot \left(IPA_{it} + \sum_{j=1}^{J} adj_{ji} \cdot IPA_{jt} \cdot ps_{ji} \right) \quad \forall i, t \qquad (8.b)$$

(9) Determining the bonus associated with harvesting blocks during period 1: the following three constraints ensure that the burn probability reduction bonus associated with harvesting block i during period 1 is the reduction in its burn probability from $Initial\_BP_i$ to 0. There is no bonus associated with any of the blocks that are not harvested during period 1.

$$Bonus_{i1} \leq X_{i1} \quad \forall i \qquad (9.a)$$
$$Bonus_{i1} \geq Initial\_BP_i - (1 - X_{i1}) \quad \forall i \qquad (9.b)$$
$$Bonus_{i1} \leq Initial\_BP_i \quad \forall i \qquad (9.c)$$

Harvesting a block during period 1 precludes it from burning during period 1. Constraint (7.a) specifies $Initial\_BP_i$, the initial burn probability for block i at the start of the planning horizon before any harvesting has taken place. If however, block i is harvested in period 1 it cannot burn during period 1. The probability that it would have burned had it not been harvested can therefore be viewed as a bonus associated with harvesting it in period 1. The larger the $Initial\_BP_i$ the larger the bonus associated with harvesting block i in period 1.

(10) Determining the bonus associated with harvesting blocks in periods $t = 2, 3, \ldots, T$: consider any block i that is harvested in period 2, ..., T. As was the case with period 1 harvests, that block cannot burn during period t so there is a burn probability reduction bonus equal to the reduction in the burn probability of block i in period t from BP_{it-1} to 0. The following three constraints determine the bonus associated with harvesting block i in period t.

$$Bonus_{it} \leq X_{it} \quad \forall i, t \geq 2 \qquad (10.a)$$
$$Bonus_{it} \geq BP_{it-1} - (1 - X_{it}) \quad \forall i, t \geq 2 \qquad (10.b)$$
$$Bonus_{it} \leq BP_{it-1} \quad \forall i, t \geq 2 \qquad (10.c)$$

2.5. Modelling the harvesting re-scheduling process

Our model can be used to generate an optimal solution to the deterministic problem that must be solved at the start of period 1. At the end of period 1 the manager will survey what has been harvested, burned and grown during the first period and use that information to update the state of the forest at the start of the second period. That updated state will then serve as input data for a slightly revised version of our planning model (one for which the planning horizon would be reduced by one period) that can be solved to re-plan what should happen during the remaining $T - 1$ periods of the finite planning horizon. We had only to add a constraint to preclude the harvesting of any cutting blocks adjacent to any blocks that had been burned during the previous period. This last constraint was included to clarify the effect of fire uncertainty on harvesting decisions. Similar state re-assessments and re-planning will take place at the end of the third, fourth … and $T - 1$ periods.

2.6. A basic model that does not account for uncertain fire loss

We will compare our approach with a deterministic approach which does not consider burn probabilities when planning. The model we choose to use for that purpose is what we describe as our basic deterministic spatial planning model — essentially our bonus model without the bonus term. Its objective function is therefore our bonus model's objective function with the bonus term removed. Given the removal of the bonus term it includes constraints (2), (3), (4), (5.a), (5.b), (6.a) and (6.b) but not constraints (7.a), (7.b), (7.c), (8.a), (8.b), (8.c), (9.a), (9.b), (9.c), (10.a), (10.b) or (10.c).

2.7. Model testing

Our mathematical programming models were solved using LINGO 10.0 with an MS Excel interface that made it possible for us to manage the input data and to evaluate the 50 simulated scenarios generated for each of the two (homogeneous and heterogeneous) landscapes studied. The CPU time in an Intel Core 2 Quad processor with 2.66 GHz and 4 GB of RAM fluctuated among 6.7−24.7 seconds per run.

Each of the 50 simulated scenarios was composed of a run of our bonus model for the entire 50-year planning horizon (the scheduling phase) and four subsequent runs of the re-scheduling versions of the model for the remaining planning periods (the re-scheduling phases) taking simulated fires into account. This sequence was repeated 50 times for the homogeneous landscape (Figure 1) and 50 times for the heterogeneous landscape (Figure 2), to determine how well the model performed when it was run for 50 different scenarios or replications, considering fire occurrences simulated for each of the five planning periods. To evaluate our inclusion of a bonus term in our objective function on harvesting decisions, all those 50 replications were run with both the basic 'no bonus' model and our bonus model.

3. Results

3.1. Levels of harvesting and terminal volume

The results show the effect of fire events on the volume of timber harvested over the whole planning horizon. The optimal total volume harvested as a result of fires that occurred

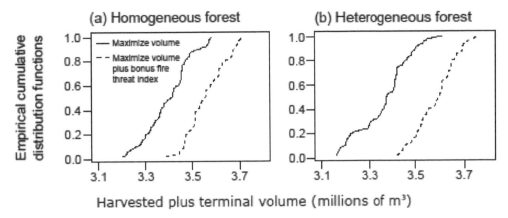

Figure 5. Empirical cumulative distribution functions for the volume harvested plus the terminal volume available to harvest when maximizing volume or maximizing volume plus the bonus fire threat index in homogeneous and heterogeneous forests.

during the planning horizon is always less than the total volume predicted when fire is ignored during the initial deterministic planning phase. That happens whether or not a bonus term is included in the objective function. However, the simulated volumes produced (both the volumes harvested and the terminal volumes left on the landscape at the end of the planning horizon) increase when the bonus is included in the objective function (see Figure 5). That difference is more pronounced when the forest has a heterogeneous structure and we believe that is because FireSmart harvesting can be focused on areas where fire threats are not concentrated and the spatial constraints are more easily satisfied.

The results presented in Tables 3 and 4 indicate that the total volume harvested increases slightly when a bonus term is used but this difference is augmented when we include the terminal volume on the landscape at the end of the planning horizon. We found increases of 4.8% and 6.5% for the homogeneous and heterogeneous forest landscapes, respectively.

One of the most significant impacts of including the bonus term in the objective function is its effect on the terminal volume growing on the landscape at the end of the planning horizon. We assume cutovers and burned stands are not regenerated so the only volume growing on the forest at the end of the planning horizon is the volume growing in stands that have not been burned or harvested before the end of the planning horizon.

Table 3. Average total volume harvested and left unburned at the end of the planning horizon with and without including the bonus fire threat index included in the objective function for our homogenous forest landscape.

	Objective function that maximizes volume alone	Objective function that maximizes volume and bonus
Volume harvested over the planning horizon (millions of m³)	2.34	2.35
Volume left at the end of the planning horizon (millions of m³)	1.07	1.21
Total volume available (millions of m³)	3.40	3.56
Volume difference (millions of m³)		0.16
Percentage difference		4.8%

Table 4. Average total volume harvested and left unburned at the end of the planning horizon with and without including the bonus fire threat index included in the objective function for our heterogeneous forest landscape.

	Objective function that maximizes volume alone	Objective function that maximizes volume and bonus
Volume harvested over the planning horizon (millions of m³)	2.36	2.39
Volume left at the end of the planning horizon (millions of m³)	1.01	1.20
Total volume available (millions of m³)	3.37	3.59
Volume difference (millions of m³)		0.22
Percentage difference		6.5%

Figures 6 and 7 indicate that for most of the simulated scenarios, the terminal volume growing on the landscape at the end of the planning horizon is larger when the bonus term is included in the objective function than when it is not.

In fact, the results show that the inclusion of a bonus term in the objective function increases the terminal volume by 14% and almost 19%, for the homogeneous and heterogeneous landscapes, respectively, in comparison with the case where the objective maximizes total volume alone. Additionally, the distribution of ignition probabilities on the heterogeneous landscape depicted in Figure 2 leads to a greater increase in the terminal volume as compared with the case when ignition probabilities are grouped more uniformly in the homogeneous forest depicted in Figure 1. This may be because it is easier to focus on harvest blocks that are at a higher fire risk in such forests, since the adjacency constraints reduce the number of cutting blocks available to harvest when those blocks are grouped in a homogeneous structure. On the contrary, an irregular distribution of fire risk (similar to Figure 2), gives more options to choose cutting blocks to harvest with spatial limitations.

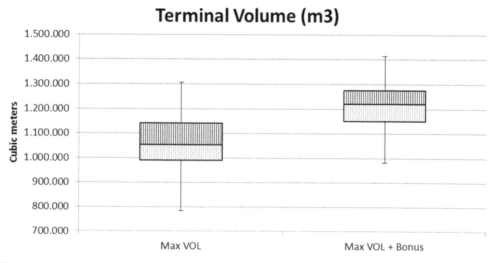

Figure 6. Distribution of the terminal volume growing on the landscape at the end of the planning horizon by objective function for a landscape with a homogeneous structure.

Terminal Volume (m3)

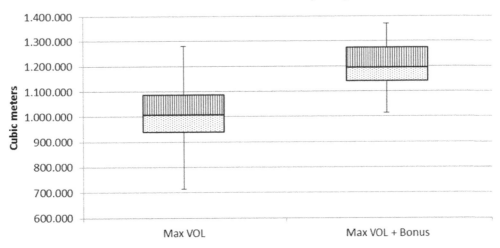

Figure 7. Distribution of the terminal volume growing on the landscape at the end of the planning horizon by objective function for a landscape with a heterogeneous structure.

3.2. Fire and area burned measures

These results are interesting with respect to the number of effective fires (i.e. blocks that actually burned) when the bonus term was included in the objective function. Figure 8 shows the distribution of total area burned due to fires by both ignition and spread by objective function for the homogenous landscape. It is clear that the inclusion of the bonus term is effective because it reduces the mean of the total area burned from 1906 to 1586 hectares. Figure 9 presents the total area burned for the heterogeneous forest and illustrates the impact of the bonus term on the area burned – reducing its mean from 1842 to 1512 hectares.

Burned Area (ha)

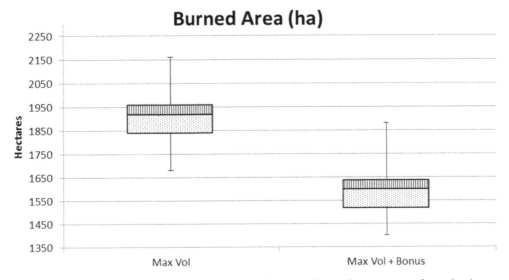

Figure 8. Total area burned by scenario and objective function for our homogeneous forest landscape.

Burned Area (ha)

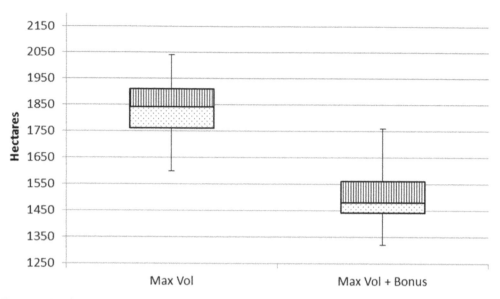

Figure 9. Total area burned by scenario and objective function for our heterogeneous forest landscape.

We can also examine the level of effective fires for each objective function and under the two different forest structures (see Figure 10). To calculate the index of effective fires we use the total number of fires simulated by ignition or spread for each scenario and the total number of burned cutting blocks (i.e. cutting blocks burned because they were not

Levels of Effective Fires

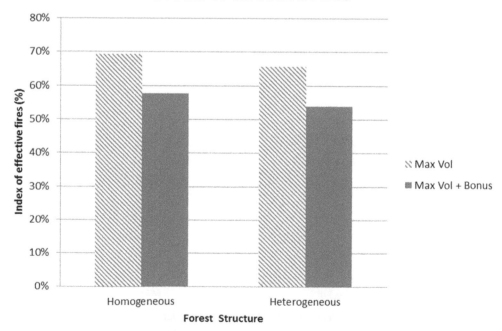

Figure 10. Levels of effective fires by objective function and forest landscape structure.

harvested), in the whole planning horizon. In this sense, for both kinds of forest, the use of the bonus reduces the number of cutting blocks that are burned. In the case of the homogeneous landscape, the level of possible fires is almost 70% when the bonus is not included and it decreases to a 58% when the bonus term is incorporated in the objective function. For the heterogeneous forest, the levels for possible fires decrease from 66% to 54% when the bonus term is included in the objective function. This indicates that when ignition probabilities are distributed irregularly on the landscape, the level of avoided fires is even less when the objective is to maximize the total volume, using a bonus term to improve the harvesting decisions for the reason explained in Section 3.1.

4. Discussion and conclusions

We developed and evaluated a new harvest scheduling model that is designed to maximize the total volume harvested from a flammable forest landscape subject to adjacency constraints over a finite planning horizon. Our approach entails the inclusion of a fire bonus term or threat index in the objective function to steer harvesting towards stands that are at greatest risk due to fire. Although we used a very simple planning problem to demonstrate its use, we believe it can and should be applied to real forest management units for which there are more realistic objectives and constraints.

Our results indicate that when such a bonus term is included in the objective function, the total volume harvested from our hypothetical forest landscapes over the planning horizon can increase by more than 1%, the terminal volume remaining unburned on the landscape can increase by almost 19% and the total area burned can decrease by 18%, in comparison with the case when a bonus term is not incorporated in the model. Not surprisingly, the structure of the landscape appears to have a significant impact on those results. When the ignition probabilities are concentrated in homogeneous groups, the impact of the bonus term is somewhat reduced in comparison with the case when more flammable stands are distributed heterogeneously across the landscape.

Our primary objective was to develop a bonus or threat index and illustrate how it can be used to help determine how best to manage timber production on flammable forest landscapes. We demonstrated our methodology by applying it to a simple hypothetical forest management planning problem with spatial attributes. We implicitly assumed fires burn for short periods by assuming they can spread no further than the cells that are immediately adjacent to the cells in which they ignite in order to simplify our burn probability constraints. Our model could easily be extended to deal with longer duration fires that spread beyond the cells that are immediately adjacent to the cell in which a fire is ignited by replacing our simple burn probability constraint with Wei et al.'s (2008) more complex burn probability constraint (8). We used a re-planning approach to apply our model to a finite five-period planning horizon problem but our approach can readily be applied to much longer planning horizons using re-planning within a rolling planning horizon context. We wanted our hypothetical planning problem to have some spatial features and focused on adjacency which in retrospect, might not be crucial for finite planning horizon problems but would be important in infinite planning horizon contexts where wildlife habitat concerns are crucial.

Acknowledgments

This research was partially funded by the Complex Engineering Systems Institute (ICM: P-05-004-F, CONICYT: FBO16) and Fondecyt grant 1120318. David Martell's contribution was funded in part by a Natural Sciences and Engineering Research Council of Canada Discovery grant. We thank LINDO Systems for the use of their LINGO software.

Disclosure statement

No potential conflict of interest was reported by the authors.

Funding

This work was supported by the Complex Engineering Systems Institute [grant number ICM: P-05-004-F], [grant number CONICYT: FBO16]; Fondecyt [grant number 1120318] and by the Natural Sciences and Engineering Research Council of Canada Discovery grant.

References

Acuña MA, Palma CD, Cui W, Martell DL, Weintraub A. 2010. Integrated spatial fire and forest management planning. Can J For Res. 40:2370–2383.

Armstrong GW. 1999. A stochastic characterization of the natural disturbance regime of the boreal mixed wood forest with implications for sustainable forest management. Can J For Res. 29:424–433.

Armstrong GW. 2004. Sustainability of timber supply considering the risk of wildfire. For Sci. 50:626–639.

Barahona F, Weintraub A, Epstein R. 1992. Habitat dispersion in forest planning and the stable set problem. Oper Res. 40:S14–S21.

Bettinger P. 2009. A prototype method for integrating spatially-referenced wildfires into a tactical forest planning model. Res J For. 3:8–22.

Bettinger P. 2010. An overview of methods for incorporating wildfires into forest planning models. Math Comput For Nat Resour Sci. 2:43–52.

Boychuk D, Martell D. 1996. A multistage stochastic programming model for sustainable forest-level timber supply under risk of fire. For Sci. 42:10–26.

Carvajal R, Constantino M, Goycoolea M, Vielma JP, Weintraub A. 2013. Imposing connectivity constraints in forest planning models. Oper Res. 61:824–836.

Chung W. 2015. Optimizing fuel treatments to reduce wildland fire risk. Curr For Rep. 1:44–51.

Cruz JP, Honeyman LP, Pezo CA, Schulze C. 2007. Growing analysis of old lenga trees (*Nothofagus pumilio*) in forest at XII Region, Chile. BOSQUE. 28:18–24.

Ferreira L, Constantino MF, Borges JG, García-Gonzalo J. 2015. Addressing wildfire risk in a landscape-level scheduling model: an application in Portugal. For Sci. 61:266–277.

Gassmann HI. 1989. Optimal harvest of a forest in the presence of uncertainty. Can J For Res. 19:1267–1274.

Goycoolea M, Murray A, Barahona F, Epstein R, Weintraub A. 2005. Harvest scheduling subject to maximum area restrictions: exploring exact approaches. Oper Res. 53:490–500.

Hirsch K, Kafka, Tymstra V, McAlpine C, Hawkes R, Stegehuis B, Quintilio H, Gauthier S, Peck K. 2001. Fire-smart forest management: a pragmatic approach to sustainable forest management in fire-dominated ecosystems. For Chron. 77:357–363.

Kim YH, Bettinger P, Finney M. 2009. Spatial optimization of the pattern of fuel management activities and subsequent effects on simulated wildfires. Eur J Oper Res. 197:253–265.

Konoshima M, Albers HJ, Montgomery CA, Arthur JL. 2010. Optimal spatial patterns of fuel management and timber harvest with fire risk. Can J For Res. 40:95–108.

Lockwood C, Moore T. 1993. Harvest scheduling with spatial constraints: a simulated annealing approach. Can J For Res. 23:468−478.

Minas JP, Hearne JW, Martell DL. 2014. A spatial optimisation model for multi-period landscape level fuel management to mitigate wildfire impacts. Eur J Oper Res. 232:412−422.

Murray AT. 1999. Spatial restrictions in harvest scheduling. For Sci. 45:45−52.

Navon DI. 1971. Timber Ram... a long-range planning method for commercial timber lands under multiple-use management. Albany (CA): USDA Forest Service, Pacific Southwest Research Station. (Research Paper PSW-RP-070).

Omi PN, Murphy JL, Wensel LC. 1981. A linear-programming model for wildland fuel-management planning. For Sci. 27:81−94.

Rand G. 2006. IFORS' operational research hall of fame − Egon Balas. Int Trans Oper Res. 13:169−174.

Reed W, Errico D. 1986. Optimal harvest scheduling at the forest level in the presence of risk of fire. Can J For Res. 16:266−278.

Rytwinski A, Crowe KA. 2010. A simulation-optimization model for selecting the location of fuel-breaks to minimize expected losses from forest fires. For Ecol Manag. 260:1−11.

Savage DW, Martell DL, Wotton BM. 2010. Evaluation of two risk mitigation strategies for dealing with fire-related uncertainty in timber supply modelling. Can J For Res. 40:1136−1154.

Snyder S, ReVelle C. 1996. Temporal and spatial harvesting of irregular systems of parcels. Can J For Res. 26:1079−1088.

Van Wagner CE. 1983. Simulating the effect of forest fire on long-term annual timber supply. Can J For Res. 13:451−457.

Wei Y, Rideout D, Kirsch A. 2008. An optimization model for locating fuel treatments across a landscape to reduce expected fire losses. Can J For Res. 4:868−877.

Weintraub A, Murray AT. 2006. Review of combinatorial problems induced by spatial forest harvesting planning. Discrete Appl Math. 154:867−879.

Yoshimoto A, Konoshima M, Marušák R. 2010. Spatially constrained harvest scheduling for strip allocation and biodiversity management. FORMATH. 9:153−172.

A model approach to include wood properties in log sorting and transportation planning

Gert Andersson, Patrik Flisberg, Maria Nordström, Mikael Rönnqvist and
Lars Wilhelmsson

ABSTRACT

There is a trend that sawmills are more focused on particular valuable products in their production. This has led to an increased demand for sawlogs that are better adapted to the target products and production efficiency. Depending on the product being produced there are different log properties which are better adapted for certain products than others. Sawmills can require hard constraints on log properties such as length, diameter, internode length and sound knots. Some properties are not required but are desired as they make the production more efficient or increase the frequencies of preferred products. In these cases, we include an added value corresponding to what the industry is willing to pay for improved adaptation of the raw material. To achieve this, we propose an optimization model that integrates logging operations (bucking and forwarding) at harvest areas, transportation planning and flexible description of demand at sawmills. High flexibility for sorting at harvest areas may require additional piles of different properties to be generated. Instead of using a large number of special assortments, we allow many sorting alternatives depending on the requirements used at the industries. The transportation planning decides on the flows between harvest areas and sawmills while considering demand and supply. Even if many potential piles are used in the planning model, only a few may be used in practice. We present computational results based on 16 synthesized geographically distributed harvest areas, each representing all regional variation of mature sample trees from the Swedish National Forest Inventory and a number of sawmills.

1. Introduction

The efficiency and the quality of production at a sawmill depend on the properties of the sawlogs delivered. Sawmills typically do not have the same property requirements as they are focused on different products, e.g. window frames, beams, construction boards and doors. Each of these products 'prefer' different 'suitable' sawlogs as raw material. Preference can be expressed in terms of 'must be' and 'would be good'. This description can be divided into one class of hard and explicit requirements ('must be') and one class where we prefer, with some evaluation, some properties ('would be good'). In this paper, we focus on sawlogs delivered to sawmills. However, the approach would be similar for pulpwood delivered to pulp and paper mills. The planning period is tactical, i.e. medium term, where we consider the destination of logs from harvest areas to industries (D'Amours et al. 2008). Sawmills commonly order sawlogs without specific quality requirements in standard assortments. One such standard assortment is, for example, pine sawlogs. This means that the sawmill has to accept and make the best use of all sawlogs delivered. Having said that, we do note that sawmills can specify that the bucking at harvest areas be prioritized by particular dimensions, e.g. specified diameter and length classes. This can be achieved by using specific bucking instructions in the bucking computers used in the cut-to-length (CTL) harvesters (machines that fell, delimb, measure and cut each stem into logs) engaged. However, specific bucking control may limit the flexibility of the overall logistic planning as there is a need to link specific harvest areas and sawmills.

The logistic planning involves selection of harvesting areas, settings for bucking control, harvesting and forwarding operations, and transportation to customers. Supply and demand is typically expressed in general assortments, such as, pine sawlogs or spruce pulpwood. Such integrated planning problems have been identified as an open problem in the forest industry, see Rönnqvist et al. (2015). If a sawmill was to ask for specific properties of the sawlogs, it would be possible to include an additional assortment or economically expressed preferences with the given specification. These possibilities may improve the integration of log processing in the forest and the manufacturing of wood products in the sawmill and further on down the value chains. However, when sawmills have different requirements, it results in an increased number of assortments in the forests. In addition, each assortment may only be adapted for use at a unique sawmill, and if so, also complicate the transportation planning. The result may lead to a decrease in flexibility (less possible destinations) in the transportation planning where the supply and demand could be difficult to balance.

The properties of sawlogs can be assessed in different ways. Individual tree information from high resolution inventories prior to harvesting, e.g. measured sample trees from the Swedish National Forest Inventory or high resolution laser scanning (Næsset et al. 2004) can be used to simulate bucking operations, i.e. cut the stems into required and desired log dimensions and assortments (Möller et al. 2003). By the use of models for predicting wood properties as knot types and sizes (Moberg 2006; Moberg & Nordmark 2006; Moberg et al. 2006), internode lengths, i.e. distances between branch whorls (Wilhelmsson 2006), basic density and heartwood (Wilhelmsson et al. 2002), the frequency of expected solid wood products can be forecasted. When it comes to the actual CTL-harvesting within a specific harvesting area (one physical site and machine set up for operation) destination, selection, bucking and sorting instructions can be supported by the bucking simulations even though the real production will be controlled by length and diameter measurements in combination with, functions estimating sound knots while crooks, and visible stem faults are judged by the harvester operator.

An alternative using additional assortments based on desired log characteristics is to make classes within the basic assortments produced at a harvesting area. However, when more than one destination is serviced from a single harvesting area, it may then be necessary to sort the logs into different piles. For example, the logging operators could sort the basic assortment 'pine sawlogs' into three new assortment piles based on three different diameter classes. In our case, we could generate e.g. 'pine sawlogs, 14−20 cm', 'pine saw-logs, 21−40 cm' and 'pine sawlogs, over 40 cm'. Each pile can be used to satisfy a demand of 'pine sawlogs' but they may also introduce more flexibility to satisfy a demand where the requirement is a more limited diameter class. A problem with this approach is that we may end up with many additional piles if many specific assortments are used. Additional piles result in higher costs in the harvesting, forwarding and transportation operations. The harvester needs to produce additional piles in the harvest area. The forwarder may have to use additional routes to pick up all piles and then sort them into additional piles adjacent to the forest roads. The logging truck may have to do more pickups of smaller piles to get a full truck load. In summary, it is beneficial to handle additional piles only if the total value chain revenues exceed the additional costs and the supplier gets sufficient economic coverage for the operations.

Although we integrate different parts, we still use local optimization for the bucking operations and the forwarding operations. An important question is how to model the transportation problem. When using specific required assortments and handling them as separate supply units, we can formulate the problem as a linear programming (LP) model. Such a model does not need to include the specific properties qualifying a log for the assortment. If we apply different sorting rules, each pile can be described with a set of properties. This problem is similar to the production planning at petroleum refineries where pooling tanks are used to describe the crude distilling units or the blending of components. Pooling models are nonlinear and non-convex and are much more difficult to solve as compared to LP models. Often there is a need to use heuristic solution approaches without any guarantee of optimality. Our problem would be comparatively large as each harvest area would represent one refinery.

In this paper, we propose an optimization model which is linear and we reduce the number of piles sorted in practice by selecting sorting alternatives which balance the number used in an efficient way. The nonlinearity is replaced by integer 0/1 variables which make the problem a mixed integer programming model (MIP). The model includes both hard constraints on the required properties, and a flexible valuation of preferred, but not required, properties. In practice, each sawmill can state their preference and an evaluation of their willingness to pay extra for improved properties. We study several forest operations including harvesting, bucking, forwarding and transportation in relation to sawmill orders of sawlogs. A general and recent description of supply chain planning in the forest product industry is found in D'Amours et al. (2008). The standard approach is to solve each of the planning problems separately. Support for forwarding operations can be found in Flisberg and Rönnqvist (2007) where routes are found by solving a routing problem. Transportation planning on a tactical level including destination planning can be found in Forsberg et al. (2005). The use of back-hauling in forestry with case studies is further described in Carlsson and Rönnqvist (2007). Operational routing of logging trucks is described in Andersson et al. (2008). There are less papers that deal with integrated planning. One example is Troncoso et al. (2015) where harvesting, transportation, production and sales planning are coordinated. Integration of harvesting and road maintenance is described in Flisberg et al. (2014). One of the main issues is the size of the resulting problems and the possibility of solving them within a reasonable length of time. In this article, a time aggregation scheme is used to improve the solvability. Bucking decisions can be integrated into the decisions made on which stands to harvest (Epstein et al. 1999). The integration of sawmill production, transportation and bucking operations is described in Maness and Adams (1993) and Lidén and Rönnqvist (2000). Some work on the integration of sorting in forests and transportation is described in Carlgren et al. (2006) who based the sorting on combination of species and parts of the tree. More references for opportunities and problems in sawmill production can be found in D'Amours et al. (2008). Most of these problems are on an aggregated and tactical level. There is a lack of articles that give in-depth details on qualities of the demand description.

As discussed above, when introducing qualities or properties in the sorting, the problem is very close to the pooling problem. The problem at refineries becomes nonlinear because properties are unknown when several products or crude oils are mixed together (see for example Misener & Floudas 2009). The pooling problem is also non-convex and solution methods cannot guarantee an optimal solution. There are different approaches based on general nonlinear solvers and special methods (see for example Misener et al. 2010). In our problem, we mix different numbers of logs and the properties can be modelled as in the pooling problem. A problem in our application is that there are many 'pooling tanks' required, one for each pile of logs. However, we can restrict the sorting possibilities based on implicit practical considerations. A consideration could be that the sorting is based only on diameter and length and implicit on the properties. This is because it is possible to describe some quality properties through, e.g. age, location and diameter. With this approach, it is possible to model it as an MIP problem. In addition, this problem has an underlying model structure we can use in the solution approach.

The main contributions of this paper is to propose an optimization model for a logistic system including harvesting, forwarding, transportation and demand at sawmills that is kept small in dimensions, a solution method that can efficiently take into account the wood properties, and a case study that evaluates the proposed model with real data. The outline of the article is as follows. First, we describe the overall problem and the different parts in Section 2. In Section 3, we propose and describe the optimization model. In Section 4, we describe a virtual case study and related analysis. The main purpose of the case study is to test the model usability and provide some generalizable examples of what the consequences may be when introducing some arbitrary hard and soft constraints on the basic log assortments; and an attempt to add explicit valuations of log properties on a limited proportion of the delivered assortments. A discussion on the proposed approach and conclusions are given in Section 5.

2. Problem description

Traditionally, procurement planning is based on a few specified assortments. The definition of an assortment is often based on species, such as pine, spruce and birch, and end usage, e.g. pulpwood, sawlog and forest fuel. We note that roundwood (including both sawlogs and pulplogs) and forest fuel assortments are, in general, not planned together. The main reason is that there are two different truck fleets used, and an additional operation, comminution or chipping is used for forest fuel. In this article, we focus on roundwood for use at sawmills. When making a tactical transportation planning, we need information on supply in harvest areas, demand in industries, transportation capacities and transportation costs. The supply at each harvest area may only be described by the volume of each assortment and the demand at each sawmill specified as a volume of an assortment group. An assortment group can be viewed as a set of assortments that can be used to satisfy a general demand. An assortment group is often a single assortment. However, sometimes several assortments can be used to satisfy a demand. For example, spruce and pine sawlogs can all be sawn if both species are accepted by the sawmill in the planning process. Today, sawmills often ask for customized price lists or apportionment matrices to be used in the bucking operations. The reason is to focus the bucking optimizer (in the harvester) to produce sawlogs with more specific lengths or diameters. In such cases, there is a pre-specified linkage between the harvest area and the industry. In forestry, there are several alternative ways to measure volume. To simplify the description, we will simply use cubic meters (m^3) as the volume measure.

We want to add the possibility of specifying more explicit requirements on the log properties delivered to the industry. There are two main aspects or possibilities with this description. First, we have *hard* constraints on property requirements. Examples of such constraints are:

- The average length of the logs should be at least 4.6 m.
- At least 80% of the logs should have a length within the interval 4.2−4.8 m.
- At least 75% of the logs should have a diameter $\geq$ 210 mm.
- At least 5% of the logs should have sound knots.

Second, we have *soft* constraints where there is an *added value* when properties are within desired threshold limits. Examples of such constraints are:

- Increased value of 40 SEK/m^3 (SEK: Swedish monetary unit kronor) if the logs have sound knots. This is valid up to a maximum of 10% of the volume.
- Increased value of 7 SEK/m^3 for each cm that the average internode length is above the 15.9 cm limit. This is valid up to a maximum of 15% of the volume.
- Increased value of 200 SEK/m^3 for straight butt end sawlogs with a maximum of five visible knots at log surface of which none are above 20 mm in diameter.
- Increased value of 1 SEK/m^3 for each kilogram increase of the wood density expressed as kg/m^3 of sawn products at 12% moisture contents. This is valid up to a maximum of 550 kg/m^3.

In order to meet such requirements on the demand side, it is important to be able to measure or assess the quality of the logs produced. Some properties such as length and diameter can easily be measured directly during harvest operations by the measurement system in the harvester felling head. Other 'interior' wood properties such as density, internode length and sound knots can be estimated based on the measured dimensions, weight and additional information such as tree age and stand location.

When log properties are taken into consideration, we need to add several parts to the planning model. First, we need to include constraints describing the hard property requirements at the industries when expressed in agreed orders. Second, we need to include the possibility of modelling increased added valuation of the log properties. Third, we would like to add the possibility of describing the log properties at harvest areas when available. The latter implies the need to describe how the sorting can be done and the maximum number of piles to generate. Furthermore, we also need to describe the

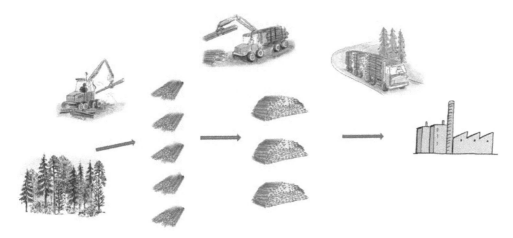

Figure 1. Description of the steps of the planning process: bucking (sorting) – forwarding (sorting) – transportation.

properties and volume of each log pile. With these additions, we no longer have a pure transportation problem. Instead, we have an integrated log operations and transportation problem with a detailed description of properties. The problem is illustrated in Figure 1. In the first step, the harvester bucks the trees and puts the logs into smaller piles. Then the forwarder collects the piles and combines them into larger piles adjacent to forest roads. The third step is when a logging truck moves the logs to an industry, for example a sawmill. Each part of the process will be discussed in more detail in the next sections.

2.1. Harvest areas

The harvester fells the tree and bucks it into logs. The bucking process is supported by an optimization routine that maximizes the value of each tree based on a price list. The price list provides a table of prices/values for each possible combination of length and diameter. The possible values for diameters are typically calculated in centimeters (cm) and the length in decimeters (dm). Figure 2 illustrates a production of logs at a larger harvest area given a price list. From the number of logs produced for different combination of length and diameter, it is clear that the production target numbers for length classes are 50, 55 and 56 dm with no focus on any particular diameter class.

Once the stems are processed, logs are positioned in different piles at the harvest area. If it is not possible to efficiently handle them together with other assortments, it is important that logs from specific assortments be put into assorted piles as they will be pooled into larger piles adjacent to the forest road once the forwarder has picked them up. Figure 3 gives an example of the distribution of logs at a harvest area after the bucking.

Given the piles produced by the harvester, the forwarder picks up the piles and moves them to the larger piles adjacent to the forest road. Here, it is important that the forwarder knows how to merge piles together, i.e. the sorting rules. The sorting can be based on many rules and approaches. The standard approach when no measured properties are used is to simply sort into a set of standard assortments as discussed earlier. However, when we consider more detailed properties and need to sort into more than one pile, we need to decide which logs are to be placed together.

	48	49	50	51	52	53	54	55	56
12	9	11	219	58	5	1	2	8	151
14	27	37	505	94	31	8	33	252	361
16	46	53	518	62	102	51	71	627	515
18	44	134	551	66	165	60	115	1189	616
20	67	188	1241	131	164	130	143	1646	1564
22	30	86	765	103	155	63	151	1692	1530
24	19	72	443	73	156	64	175	1914	1056
26	65	99	605	63	152	84	126	1739	926
28	23	129	525	42	130	88	125	1630	442
30	6	18	244	23	86	57	80	1364	517
32	4	18	273	43	73	79	73	1122	322
34	4	7	6	10	101	35	53	957	191

Figure 2. An example of a production outcome (number of logs) using a given bucking instruction. The length (horizontal axis) of the produced logs is in dm and the diameter (vertical axis) is in cm.

Figure 3. An example of log piles distributed at a harvest area. The larger square to the right denotes the position of piles adjacent to the forest road. Each smaller square represents one pile and the lines represent the trail network available to the forwarder.

Given information on the harvest area and information from the bucking (of each log), we can compute the properties of all potential sorted piles. Some of the properties we can assess for each log are: length, diameter, sound knots and internode length (i.e. average distances between branch whorls). We note that this is just a small number of possible properties. We address only a small number in this article to make the description more clear. Some properties, like length and diameter, can be measured directly by the harvester and can be used as a criterion for sorting. Other properties are currently not determined by the harvester, but there are statistical models available to determine these properties based on harvester data.

In the proposed model, we assume that all properties measured can be used as a basis for sorting. Since the model will be used as a tactical tool where bucking information is obtained through simulations, we can use different bucking instructions on the model trees in order to produce a number of outcomes for which we can compute log properties. Once we have bucked all trees, we have detailed information on each log produced based on combinations of diameter and length. Given this data, we can combine suitable combinations of diameter and lengths into piles. We illustrate this approach with an example of how to implement sorting rules and compute necessary information. For simplicity, we only consider diameter and length as properties. In Figure 4, we provide the average length (l), average diameter (d) and volume (v) produced for each combination of diameter and length.

Given the information in Figure 4, we can define different sorting rules. In Figure 5, we illustrate a case with four sorting alternatives. In each sorting alternative, we produce a set of piles and each corresponds to a number. In the first alternative (top left), we only produce one pile as we put all produced logs into one pile. In this pile, we have 407 m^3 with an average length of 5.11 m and an average diameter of 24.6 cm. In the second alternative,

	4.0–4.5 m	4.5–5.0 m	5.0–5.5 m	5.5–6.0 m
15–20 cm	l=4.2 m d=19 cm v=20	l=4.7 m d=18 cm v=25	l=5.2 m d=17 cm v=30	l=5.7 m d=16 cm v=27
20–25 cm	l=4.2 m d=24 cm v=15	l=4.8 m d=23 cm v=20	l=5.4 m d=22 cm v=30	l=5.7 m d=21 cm v=28
25–30 cm	l=4.4 m d=28 cm v=30	l=4.9 m d=27 cm v=35	l=5.4 m d=27 cm v=33	l=5.8 m d=26 cm v=35
30–35 cm	l=4.5 m d=32 cm v=17	l=4.9 m d=33 cm v=22	l=5.3 m d=34 cm v=24	l=5.6 m d=32 cm v=16

length (m)

l : length

d : diameter

v : volume

diameter (cm)

Figure 4. An example of a set of logs produced according to a given bucking instruction. Each combination of length and diameter class provides a volume and an average diameter and length.

1	1	1	1
1	1	1	1
1	1	1	1
1	1	1	1

Sorting 1: length = 5.11, diam = 24.6, vol = 407

1	1	2	2
1	1	2	2
1	1	2	2
1	1	2	2

Sorting 1: length = 4.61, diam = 25.6, vol = 184
Sorting 2: length = 5.51, diam = 23.8, vol = 223

1	1	2	2
1	1	2	2
1	1	3	3
3	3	3	3

Sorting 1: length = 4.58, diam = 23.7, vol = 145
Sorting 2: length = 5.49, diam = 19.0, vol = 115
Sorting 3: length = 5.32, diam = 29.9, vol = 147

1	1	2	2
1	1	2	2
3	3	4	4
3	3	4	4

Sorting 1: length = 4.51, diam = 20.6, vol = 80
Sorting 2: length = 5.49, diam = 19.0, vol = 115
Sorting 3: length = 4.69, diam = 29.4, vol = 104
Sorting 4: length = 5.54, diam = 29.0, vol = 108

Figure 5. Illustration of four sorting alternatives. For each alternative, we provide the average diameter, average length and volume when the diameter and length classes from the price list are combined. The number in each box represents which pile number it is sorted into. The first alternative (top left) produces one pile. The second (top right) produces two piles. The third (bottom left) produces three piles. The fourth (bottom right) produces four piles.

we sort into two piles. The sorting is based on a pile with shorter lengths and the other with longer lengths. The average lengths are 4.61 m and 5.51 m, respectively. In the third and fourth alternatives, we produce three and four piles, respectively. Clearly, we can satisfy strict requirements on lengths and diameters better with alternative four. However, it requires handling of more piles and results in increasing costs.

Sorting into several piles results in larger production costs. The harvester needs to evaluate and position more piles and the forwarder needs to make several runs to pick up all piles. The Forestry Research Institute of Sweden has previously developed and evaluated detailed functions to evaluate the costs for harvesting and forwarding. These costs depend on the number of piles in the chosen sorting alternative.

2.2. Industrial demand

In traditional planning, each industry, e.g. sawmills, only needs to provide a volume and an assortment to define a general demand or information for the design of bucking instructions concerning preferred combinations of log lengths and diameter. In the proposed approach, both hard constraints and soft valuations of preferred properties must be defined. We have given some specific examples of hard constraints earlier. We can extend the description to a general structure as formulated below.

- The average value of 'property x' of the logs should be at least (or at most) *y* units.
- A proportion *y*% of the logs should have the value of 'property x' within the interval [w1, w2].

The preferred values of some properties are not hard constraints. In order to meet such a preference, we must evaluate how much an increase in a property is worth. It is important to measure the increased value in the same monetary term as the production and transportation costs, e.g. SEK/m^3. Because the sawmills may define specific valuation of log properties for specific log dimensions, it is difficult to formulate a general structure for the hard constraints. Specific examples of soft constraints or valuations are:

- Increased value of 40 SEK/m^3 if the logs have sound knots. This is valid up to a maximum of 10% of the volume.
- Increased value of 10 SEK/m^3 for each centimetre that the average internode length is above the limit of 16 cm. This is valid up to a maximum of 20% of the volume.

Suppose the demand is 1000 m^3. If 5% of the delivered volume of 1000 m^3 has sound knots, the company is willing to pay 2000 SEK (40 × 0.05 × 1000). If the average internode length is 20 cm in a delivered volume of 1000 m^3, the company is willing to pay 8000 SEK (10 × (20−16) × 0.2 × 1000). In the proposed model, we represent this with two different constraints. The first constraint provides the overall demand of 1000m^3 and uses volumes. The second constraint measures the deviation from the property level of 5%. Here, the demanded volume is multiplied with the target level in the right-hand side and the volume from each pile is multiplied by the property of each pile. For example, if one pile has 10% sound knots, we can multiply this with the actual flow between the pile and sawmill. Any deviation is measured by extra variables and the objective coefficients of these additional variables can be used to compute the increased value. This technique is the same used in refinery models to measure characteristics on the petroleum products in pooling processes (Amos et al. 1997).

A way to decrease the number of alternatives and focusing on specific products is to apply explicit index valuation where important properties are combined based on estimated impact on the sawmill production and/or frequency of desired products. This is the same approach as practised in tree breeding (Alzamora & Apiolaza 2010; Ivkovic et al. 2010). As long as the logs can be extracted to different production alternatives downstream in the value chain, there is still a possibility for changes in valuation based on suitable or unsuitable properties. This flexibility will decrease as the logs are piled and destined to a specific industry but may increase again if the logs are sorted by an advanced scanner at the sawmill.

2.3. Transportation

In traditional transportation planning, we decide on flows between assortment piles and demand expressed in assortment groups. In this problem, the number of piles are predetermined. In our new transportation problem, we determine flows from potential sorted piles (with unique values of properties) to orders at sawmills. One part of the problem is deciding on which sorting alternative to use at each harvest area as discussed earlier. The transportation planning considers flows and, hence, we need to define a unit transportation cost. This is only an approximation of the real cost when logging trucks are used in the operation. The reason comes from the fact that not all piles provide an integer number of full truck loads. In the standard problem, this is not so much of a problem as any error will be about the same for any average harvest area. However, when dealing with additional separated assortments, the number of piles increases. This implies that harvest areas where sorting into more piles is carried out, may also make more piles which are not full truckloads. RuttOpt (Andersson et al. 2008) is a routing system developed at The Forestry Research Institute of Sweden to schedule truck transports to fulfil demand requirements of different assortments given a supply. We have used RuttOpt in some simulations to determine the total transportation cost for different numbers of piles/assortments at supply points by performing a number of simulations *a priori*. RuttOpt provides detailed daily routes for all trucks and provides a very detailed estimate of the total transportation cost. In the simulations, we increased the number of piles gradually and measured the increased routing cost which depends on an increased number of multiple not fully loaded truck pickups. The simulations provided an extra cost estimate of each extra pile which we have used in the proposed model.

3. Model

The optimization model consists of two main sets of decisions. The first set is to decide which sorting alternative to use at each harvest area. All sorting alternatives need to be generated and there are several possibilities. Each sorting alternative is represented with a binary decision variable. The second set is the flows between piles (generated through the sorting alternatives) and industries. We also use variables to determine the difference in values between actual and threshold levels for preferred properties. To ensure that the model can always establish feasible solutions, we include variables to measure any infeasibility. The constraints describe supply, demand, hard constraints on properties and constraints describing the preferred properties. The objective function is to balance revenue from providing desired properties while simultaneously minimizing the overall cost of logging operations (including sorting) and transportation.

Sets

I: set of supply points (harvest areas)
J: set of demand points (industries)
G: set of assortment groups
P: set of piles
P_g^G: set of piles that can be used to supply assortment group g
T: set of time periods
S: set of sorting alternatives
S_i^I: sorting alternatives available at supply point i
V_{jg}^B: set of property requirements at demand point j for assortment group g
R: represents all defined flows (i, j, p, t)

Parameter data

c_{ijp}^T: transport cost from supply point i to demand point j for pile p
c_{is}^S: cost for sorting option s at supply point i
$\bar{c}_{jgvt}^F, \underline{c}_{jgvt}^F$:   cost/bonus (negative value) if property requirement v is over/under required value
c_{jgt}^P: penalty cost for unfulfilled demand
S_{isp}:   available volume of pile p if sorting s is used at supply point i
D_{jgt}:   demand of assortment group g at demand point j in time period t
b_{jgvt}^F: property requirement level
a_{ipv}^B: level of property requirement v at supply point i of pile p

The unit transportation cost (SEK/m^3) is based on agreements between forest companies and transportation companies. This unit cost is based on a concave piece-wise linear function on the distance. The cost of sorting option is based on detailed studies from logging operations when additional assortments are produced together with the increased cost in the transportation due to several pickups of small piles.

Variables

$$x_{ijpt} = \text{flow from supply point } i \text{ to demand point } j \text{ using pile } p \text{ in time period } t$$
$$y_{is} = 1 \text{ if sorting option } s \text{ is used at supply point } i, 0 \text{ otherwise}$$
$$l^I_{ip} = \text{inventory of pile } p \text{ at supply point } i \text{ at the end of the planning horizon}$$
$$w^P_{jgt} = \text{unfulfilled demand}$$
$$d^{F+}_{jgvt}, d^{F-}_{jgvt} = \text{difference in demand property requirement (above/below) for property requirement } v \text{ compared to requested}$$

The model

$$\min z = \sum_{i,j,p,t \in R} c^T_{ijp} x_{ijpt} + \sum_{i \in I} \sum_{s \in S} \sum_{t \in T} c^S_{is} y_{is} + \sum_{j \in J} \sum_{g \in G} \sum_{t \in T} c^P_{jgt} w^P_{jgt}$$

$$+ \sum_{j \in J} \sum_{g \in G} \sum_{v \in V} \sum_{t \in T} (\bar{c}^F_{jgvt} d^{F+}_{jgvt} + \underline{c}{jgvt}^F d^{F-}_{jgvt})$$

subject to

$$\sum_{j,t:(ijpt) \in R} x_{ijpt} + l^I_{ip} = \sum_{s \in S^I_i} S_{isp} y_{is}, \quad \forall i, p \tag{1}$$

$$\sum_{s \in S^I_i} y_{is} \leq 1, \quad \forall i \tag{2}$$

$$\sum_{p \in P^G_g, i:(ijpt) \in R} x_{ijpt} = D_{jgt} - w^P_{jgt}, \quad \forall j, g, t \tag{3}$$

$$\sum_{p \in P^G_g, i:(ijpt) \in R} a^B_{ipv} x_{ijpt} = D_{jgt}(b^F_{jgvt} + d^{F+}_{jgvt} - d^{F-}_{jgvt}) \quad \forall j, g, v \in V^B_{jg}, t \tag{4}$$

$$x_{ijpt}, l^I_{is}, w^P_{jgt}, d^{F+}_{jgvt}, d^{F-}_{jgvt} \geq 0, \quad \forall i, j, p, g, v, l, t \tag{5}$$

$$y_{is} \in \{0, 1\}, \quad \forall i, s \tag{6}$$

The objective is to minimize the total costs expressed as transportation costs, sorting costs at supply points (including harvesting and forwarding costs and an approximation of increased transportation cost due of extra piles), cost of unfulfilled demand and cost/bonus for extra demand properties. Constraint set (1) describes the available supply depending on the chosen sorting option. Constraint set (2) describes that at most, one sorting alternative can be used at each supply point. Total demand must be fulfilled in constraint set (3). Both hard requirements on properties and preferred levels of properties are described in constraint set (4). The hard requirements are fulfilled by allocating high values to the parameters $\bar{c}_{jgvt}^{F}$ and $\underline{c}\,jgvt^{F}$ which correspond to unacceptable values. Non-negative variables are described in constraint set (5) and binary variables in constraint set (6).

4. Computational results

4.1. Case study

In the case study, we have information on 16,491 logs from 6941 real sample trees (both pine and spruce) extracted from the Swedish National Forest Inventory (SNFI) between the years 2001 and 2005. These trees were originally distributed to 600 geographically spread sample plots in the county of Västerbotten judged as mature for final cut. The trees were pooled together on a geographical neighbourhood basis into 16 harvest areas in order to synthesize stand-like variation in tree sizes. The total volumes available of saw-logs are 611 m^3 and 456m^3 of pine and spruce, respectively. In this analysis, we have no demand of pulpwood and it is already determined if a log will be used as sawlog or as pulpwood. Therefore, we can remove the pulpwood information from the analysis.

The demand volumes of pine and spruce sawlogs at seven different sawmills is given in Table 1. We have removed the real names and will instead use the names Sawmill1 – Sawmill7. In the case study, there is a demand for both pine and spruce sawlogs but we focus on property requirements for pine only. Since the total demand is 8801m^3, we have scaled the available wood volume by copying each tree (from the SNFI) nine times. This will still be a good description of the logs available.

Figure 6 gives the geographical area of the harvest areas and the seven sawmills used in the case study.

By using RuttOpt *a priori* as mentioned earlier, we have estimated an average unit cost of 417 SEK for each extra potential pile to be transported from a harvest area. This can be included as a fixed cost for each sorting alternative (and depending on the number of piles) in the model. The harvesting and forwarding costs are estimated using detailed functions developed by the Forestry Research Institute of Sweden. We note that the supply information is not provided from measured harvest areas. Instead, it is based on aggregated information from many small plots. The main purpose of this case is to test the characteristic of the model and its solvability by using a case with data that will represent a production environment with high variability (all local variation from sample plots are mixed into the same stand). We will, hence, not make any detailed analysis and conclusions on the actual case.

Table 1. A description of the demand volumes at the mills.

Sawmill	Assortment	Volume (m^3)
S1	Pine sawlog	2211
S2	Spruce sawlog	1194
S3	Pine sawlog	1038
S4	Spruce sawlog	571
S5	Spruce sawlog	1297
S6	Spruce sawlog	441
S6	Pine sawlog	882
S7	Spruce sawlog	389
S7	Pine sawlog	778

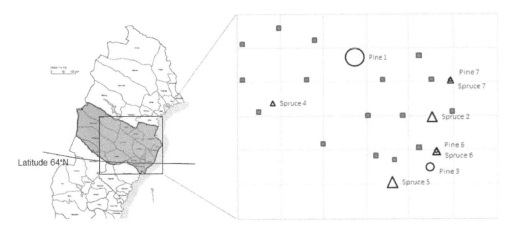

Figure 6. The county of Västerbotten, northern Sweden, is shown to the left. Dark represents the area of land included in the analyses. The square refers to the harvest areas and circles and triangles to the location of virtual sawmills shown in the right part. Symbol sizes indicate relative volume demand (spruce and/or pine) at the mills and the mill number and type of demand is written beside. Mills with demand for spruce and pine are located on top of each other.

In the case study, we have three sawmills (S1, S3 and S6) with requirements and valuations of pine sawlog properties. This information is summarized in Table 2. The type of constraints on properties is either hard constraints (*hard*) or soft increased values (*soft*). Note that the sawmill S7 requires pine sawlogs but does not have any restrictions on properties. Sawmill S2, S4 and S5 demand spruce sawlogs and there are no requirements and valuations.

The bucking simulations behind this study are controlled by a standard 'length neutral' price list (see Table 3) with length in centimeter and diameter in millimeter. This means

Table 2. A description of the restrictions/valuation of the properties used at three of the four pine-based sawmills in the case study. Predicted as wood density at 12% moisture content.

Sawmill	Property	Type	Descriptions
S1	Length	Hard	At least 80% of the logs should have a length within the interval 4.3–4.9m.
S1	Length	Hard	Max 5% of the logs should have a length of < 3.7m and > 4.9 m.
S1	Length	Hard	The average length of the logs should be ≥ 4.55 m.
S1	Diameter	Hard	At least 25% of the logs should have a top diameter within the interval 134–165mm.
S1	Sound knots	Soft	Value of 50 SEK/m³ if the logs have sound knots. Valid up to a maximum of 10% of the volume.
S1	Internode length	Soft	Value of 7 SEK/m³ for each cm the internode length is over 15.9cm. Valid up to a maximum of 5% of the volume.
S1	Density	Soft	Decreased value of 5 SEK/kg if the density is under 420 kg/m³ in the volume 'internode length'.
S3	Length	Hard	At least 90% of the logs should have a length of 4.3 m or higher.
S3	Diameter	Hard	At least 80% of the logs should have a top diameter of 160mm or higher.
S3	Length	Hard	The average length of the logs should be ≥ 5.35 m.
S3	Sound knots	Soft	Value of 55 SEK/m³ if the logs have sound knots. Valid up to a maximum of 15% of the volume.
S6	Length	Hard	The average length of the logs should be ≥ 4.8 m.
S6	Sound knots	Soft	Value of 45 SEK/m³ if the logs have sound knots. Valid up to a maximum of 15% of the volume.

Table 3. A length neutral price list (pine butt logs) used for all the bucking simulations. All prices in SEK per m³ top diameter (under bark).

Length/diam	134	140	160	180	200	220	240	260	280	300
340	329	403	439	495	516	525	530	533	537	550
370	391	428	464	520	541	550	555	558	562	571
400	406	443	479	535	556	565	570	573	577	590
430	426	463	499	555	576	585	590	593	597	606
460	438	475	511	567	588	597	602	605	609	618
490	445	482	518	574	595	604	609	612	616	625
520	458	488	526	582	603	612	617	620	620	630
550	453	491	552	582	603	612	617	620	624	633

that we do not try to produce additional logs in a particular length and diameter combination. Table 4 gives the volume proportions of different length and diameter combinations produced using the length neutral price list.

4.2. Instances

To test the proposed model under different scenarios, we have constructed eight instances, C1−C8, described in Table 5. Instance C1 is the base case where only volumes of basic assortments are required. In instances C2−C5, we sequentially include additional requirements of the properties. In instances C6−C8, we analyse the effect of changing the logging operation costs.

We use four different sorting alternatives in the analysis. We can easily add more alternatives but to keep the description illustrative we limit ourselves to these four. Each alternative generates one pile of spruce sawlogs and one or several pile(s) of pine sawlogs (and

Table 4. Volume of each length and diameter combination given in percentage of total volume using a length neutral price list.

Length/diam	134	140	160	180	200	220	240	260	280	300	Total
340	0.00	0.00	0.00	0.00	0.00	0.00	0.00	0.00	0.00	0.00	0.00
370	1.07	0.80	1.26	1.91	1.38	1.28	1.05	1.21	0.40	0.36	10.72
400	0.36	0.69	0.84	1.16	0.73	0.66	0.79	0.52	0.30	0.37	6.43
430	0.61	0.71	1.16	2.30	2.14	2.14	1.20	0.91	0.25	0.67	12.10
460	0.37	0.93	1.10	2.24	2.31	2.83	2.07	1.75	1.26	1.98	16.84
490	0.35	0.97	0.90	1.52	2.47	2.49	1.60	1.45	1.18	1.08	14.00
520	0.04	0.86	0.52	2.86	3.48	3.21	4.19	2.41	1.20	2.11	20.88
550	0.00	1.74	1.87	2.82	2.51	1.95	2.18	1.57	1.20	3.20	19.04
Total	2.80	6.72	7.65	14.80	15.01	14.57	13.08	9.82	5.79	9.77	100.00

Table 5. A description of the eight instances.

Instance	Descriptions
C1	Only volume restrictions and no evaluations and requirements of properties
C2	C1 and soft evaluation of the properties
C3	C2 and length requirements (no average length requirements)
C4	C3 and diameter requirements
C5	C4 and average length requirements
C6	C3 but logging operation costs decreased to 0.1*(logging operation costs in Case C3)
C7	C3 but logging operation costs decreased to 0.01*(logging operation costs in Case C3)
C8	C3 but logging operation costs decreased to 0

Table 6 A description of the different sorting alternatives for sawlog pine.

Sorting alternative	Description
NoSorting	Only basic assortments
Length	Pine divided into three different piles depending on length (three potential piles)
LenDia	Length sorting plus sorting alternative with three different diameter interval ($3 \times 3 = 9$ potential piles)
Complex	LenDia sorting plus combinations with all property sorting alternatives (54 potential piles)

pulpwood which is not included in the analysis). The different sorting alternatives are given in Table 6. The first alternative is no sorting at all. The second is to sort only into three different length intervals (sorting alternative 'length'). The third is to include three different diameter intervals together with the length sorting (sorting alternative 'LenDia'). The fourth sorting rule alternative 'complex' sorts the logs according to the alternative LenDia and according to 'sound knot' and 'internode length' (which is discretized in three internode lengths). Note that we may not need to use all 54 potential piles in the 'complex' sorting alternative. When we decide the transportation planning, it may result in several piles going to the same mill and, hence, they can be grouped together in the beginning. This is more a consequence of using a small number of alternatives. Although we generate a large number of alternatives, we would at most sort one pile for each sawmill.

4.3. Results

All tests have been done on a standard PC with 2.67 GHz and 1 GB of internal memory. Microsoft Excel is used for data handling and a VBA macro is used to extract relevant data for the optimization problem. The model is implemented in the modeling language AMPL and solved using the solver CPLEX version 11.0. In the case with the largest number of sorting alternatives, the model has about 100 binary variables, 3200 continuous variables and 1900 constraints. It takes less than a second to solve the problem.

If we use only volume requirements (instance C1), it is possible to find feasible solutions. However, for some combinations of sorting alternatives and instances, it is not possible to find a feasible solution. The reason is that the restrictions on properties are too hard with respect to what can be sorted at the harvest areas. In Table 7, the different instances are marked *feasible* when they could be solved with a specific sorting alternative, and 'n.f.' when no feasible solution existed. Obviously, we have more flexibility to meet requirements with more sorting as is given by alternative 'complex'.

To compare the impact of increasing requirements, we have summarized total objective function value, transportation costs, logging operations and added value for the cases with

Table 7 A description of when a feasible solution exists for the different instances and sorting strategies. When no feasible solution is found, it is indicated with 'n.f.'

Sorting alternative	C1	C2	C3	C4	C5
NoSorting	Feasible	Feasible	n.f.	n.f.	n.f.
Length	Feasible	Feasible	Feasible	n.f.	n.f.
LenDia	Feasible	Feasible	Feasible	Feasible	Feasible
Complex	Feasible	Feasible	Feasible	Feasible	Feasible

Table 8. The objective value (in 1000s) for different cases when the requirements on properties are changed. The number of required sorting of each alternative is also given.

Instance	Objective	Transportation cost	Logging operations	Added value	Number of sorting using			
					No sorting	Length	LenDia	Complex
C1	1638	635	1004	0	15	0	0	0
C2	1609	635	1004	28.9	15	0	0	0
C3	1715	703	1043	30.5	5	10	0	0
C4	1768	727	1072	30.9	3	7	5	0
C5	1835	778	1082	25.9	3	7	4	0

different demand requirements in Table 8. Included is also information on how many of the mills that require any of the four sorting alternatives. The objective function value for each instance is the same for all sorting strategies where a feasible solution exists. This is because the increase in logging operation cost for more piles is high, so the least number of piles to fulfill the demand and hard requirements on properties is always selected. In the logging operations presented in the table, we have included all production costs, i.e. harvesting cost + forwarding cost and also an estimated extra transportation cost due to extra potential piles to be transported from harvest areas (i.e. the 417 SEK per pile). It is clear that increased requirements, i.e. increased number of hard constraints, increases the overall cost. The added value is relatively low as compared to the overall cost, but this reflects the real situation. Another valuation of the added value would, of course, result in other solutions.

The same cost components (objective, transportation cost, logging operation cost and added value) are given in Table 9 when the logging operation costs are decreased. Here, we present the results with only two different sorting alternatives. As the logging operation cost corresponds to a large proportion of the overall cost, we have very different objective values for the cases. We can conclude that more flexibility, i.e. more sorting alternatives to choose from in the sorting 'complex' provides the best result.

The transportation cost per volume delivered to each mill with pine sawmill requirements changes for the different instances (see Table 10). The sorting alternative used is 'LenDia'. This transportation cost is calculated as total cost of transportation to the specific mill and assortment divided by total volume to the mill. In the same table, we also present the added value for requested properties. It is interesting to note that sawmill S7 increases its cost even though it does not introduce any restriction of the properties. The reason for this is that all integrated logistic operations to all sawmills influence each other. If one sawmill requires a particular type of log, it will also change the situation for the other. It may lower the cost or it may increase the cost. All this depends on the geographical location of the mills and the harvest areas.

Table 9 The cost value (1000s) for cases with different logging operation costs when two sorting alternatives (LenDia and Complex) are used.

Instance	Objective		Transportation costs		Logging operation costs		Added value	
	LenDia	Complex	LenDia	Complex	LenDia	Complex	LenDia	Complex
C4	1768	1768	727	727	1072	1072	30.9	30.9
C6	768	768	684	683	117	117	31.7	32.2
C7	662	660	682	684	12.0	12.6	32.2	35.8
C8	650	648	682	684	0	0	32.3	36.6

Table 10 The transportation costs (TrpCost) and added value (AddVal) for pine sawlogs for different cases.

Sawmill	C1		C2		C3	
	TrpCost	Addval	TrpCost	Addval	TrpCost	Addval
S1	68.0	0	68.0	8.5	87.4	10.0
S3	51.5	0	51.5	6.0	53.3	2.4
S6	47.5	0	47.5	4.3	62.9	6.8
S7	30.5	0	30.5	0.0	43.6	0.0

Sawmill	C4		C5		C8	
	TrpCost	Addval	TrpCost	Addval	TrpCost	Addval
S1	91.6	10.4	101.6	10.1	88.4	9.9
S3	66.2	1.9	108.3	0.6	57.2	4.7
S6	52.0	6.8	46.6	3.5	54.6	6.2
S7	57.9	0.0	55.7	0.0	43.6	0.0

Table 11 Harvesting and forwarding costs (SEK/m^3) for pine sawlogs when sorting alternative LenDia is used and two cost allocation alternatives 'even' and 'origin'.

Sawmill	C1		C2		C3		C4		C5	
	Even	Origin	Even	Origin	Even	Origin	Even	Origin	Even	Origin
S1	108.6	113.9	108.6	113.9	115.1	120.4	119.6	124.7	123	125.6
S3	108.6	99.5	108.6	99.5	115.1	105.3	119.6	109.1	123	124.1
S6	108.6	107.8	108.6	107.8	115.1	109.8	119.6	112.8	123	114.5
S7	108.6	106.3	108.6	106.3	115.1	118.9	119.6	127.1	123	123.7

The logging operation costs (harvesting, forwarding and estimated extra transportation costs due to extra potential piles) vary for the different cases. In Table 11, we compare two different alternatives of allocating the logging costs to different mills where the costs are given in SEK/m^3 and the sorting alternative is LenDia. The first alternative is to compute the total logging operation cost divided by the total demand for all mills (column 'even' in the table). The second alternative is to allocate the logging operation cost from each supply point to the mills that receive products from the supply point (column 'origin' in the table). If the mills belong to the same company, it may be possible to use an even distribution of the cost. Otherwise, it would be very difficult and each mill would need to handle their own costs. Even if the mills belong to the same company, they need to redistribute the cost as they often act as independent units. We observe the difficulty of finding a distribution which can be regarded as fair among the sawmills. There is an increasing number of articles describing cost allocation models and Guajardo and Rönnqvist (2015) provides a survey of such allocation models.

Table 12 Usage of different sorting options.

Sorting option	C1	C2	C3	C4	C5	C6	C7	C8
NoSorting	15	15	5	3	3	2	0	0
Length	0	0	10	7	7	7	4	4
Diameter	0	0	0	5	5	7	12	12

Table 13 Average value of properties of the delivered volumes of pine sawlogs for instances C1, C3, C4, C5 and C8.

Property	Case	S1	S3	S6	S7
Length	C1	4.65	4.65	4.71	4.68
(m)	C3	4.51	4.96	4.53	4.77
	C4	4.50	5.03	4.42	4.84
	C5	4.55	5.35	4.80	4.25
	C8	4.51	4.84	4.42	4.96
Diameter	C1	217.4	229.7	224.6	223.0
(mm)	C3	220.2	232.6	220.6	214.1
	C4	212.1	244.5	219.0	224.7
	C5	212.3	246.0	233.7	217.8
	C8	210.0	234.7	225.1	230.0
Sound knot	C1	0.08	0.11	0.1	0.1
(proportion)	C3	0.09	0.04	0.15	0.11
	C4	0.1	0.03	0.16	0.09
	C5	0.09	0.01	0.08	0.17
	C8	0.09	0.09	0.14	0.07
Internode	C1	6.97	8.67	6.85	7.56
length	C3	9.12	5.53	7.44	5.83
(cm)	C4	9.21	4.96	7.93	5.79
	C5	9.12	2.74	6.38	8.61
	C8	8.8	6.78	7.99	5.02
Density	C1	489.7	498.8	501.5	499.7
(kg/m^3)	C3	489.4	504.7	495.8	495.7
	C4	488.5	504.4	497.7	497.1
	C5	490.5	504.6	506.7	489.6
	C8	488.4	499.8	496.8	502.1

In Table 12, we give the number of harvest areas that different sorting options have been applied to when the sorting alternative LenDia has been used. Clearly, we gradually need more advanced sorting alternatives as we increase the requirements.

We can also study how the delivered logs changes with different scenarios. The average values of the properties (of delivered pine sawlogs) at mills are presented in Table 13. This information provides a good basis for analysing the effect of log properties on all sawmills when some of them impose hard constraints. If one mill requires some properties, the other will also be impacted. Whether it is a positive or negative change depends on how their desire is aligned with the requirements imposed. To guarantee a particular value it is necessary to include these in the specification.

5. Concluding remarks

The problem of integrating wood properties with hard and soft constraints, sorting alternatives and transportation has not previously been formulated. One of the main contributions of the paper is that we have proposed a generic approach to include this in the planning process. An important part of this approach is to formulate an optimization problem that includes both a quality description of the logs and integrates bucking, forwarding, transportation and added values at deliveries. The proposed model is an MIP model which can be difficult to solve even for small sizes. However, the solution time in our case is very short. We expect to be able to solve it efficiently even if the number of harvest areas, assortments, industries and constraints increases considerably. The reason for quick solution time depends, most likely, on the underlying network structure and the fact that one sorting alternative often is the most efficient alternative at each harvest area.

The constraints to model log properties are very flexible and general. For example, it is easy to add constraints or target values to different length and diameter classes. This is already aligned with the approach we use to define sorting alternatives.

The proposed model requires more information than a traditional sequential approach. Some information is already available in today's operations. For example, detailed information on log dimensions is already generated today by modern harvesters. Further wood properties can be estimated based on the measured characteristics. For example, given the age of the trees and location (based on harvest area information), stem diameter (1.3 m above ground) and length measurement along the stems (or taper functions for bucking simulations), we can predict many properties like basic and green wood densities, knot sizes and types, distances between branch whorls, heartwood content, ring patterns as well as fibre properties and bioenergy properties. This information can be used to calculate the best alternatives before harvesting and exclude inferior alternatives before running a model for total optimization. Furthermore, the bucking instructions can be much better adapted to different requirements than in this case where we have not worked with these possibilities. Information on sorting, forwarding and transportation is also available although we need to make estimations on the cost for extra piles. Assessing the added value for certain properties may be more difficult, since this relies on the knowledge and experience of the production planner at the sawmill. However, with such a system for valuation in place, incentives should be created to bring the most suitable wood raw materials from the forest to each sawmill.

The model can be used on a tactical or operational planning level to determine sorting strategies and destination of logs between harvest areas and sawmills. Given a sorting alternative, it is possible to implement this directly in harvest operations. This is true when the sorting is based on length and diameter combinations, as dimensions are measured by the harvester. If statistical models for other wood properties like internode length and sound knots are implemented in the bucking computers, more complex sorting alternatives can also be used directly. The model makes it possible to use many sorting alternatives, which can be defined in different ways. In this work, we have used basic rules. It would also be interesting to include and test sorting rules based on index values, which may describe the relative importance between different properties, harvest areas and sawmills. It is also possible to test the impact of different levels of added values at the sawmills.

The case in this paper is generic and we do not make any detailed conclusions on the actual simulated operations. However, we can make some general comments on the results. It is clear that an increased number of hard constraints results in higher logistic costs. Therefore, it is important to find a balance between additional hard constraints at sawmills and increased logistic costs. In our case, we did not use an added value for the hard constraints. However, it is easy to include added value for an increase in delivered logs which satisfy some hard constraints. This is a negotiation process between the forest company and each sawmill. The proposed model provides such a capability and can be a very efficient and valuable analysis tool.

The model proposes a full solution with respect to logging operations and transportation. This solution can be improved by also including different alternative bucking simulations in the model. This is a very interesting development to add into the overall solution process.

If each sawmill is responsible for paying for delivered volumes, the increased costs may affect the sawmills in an unintentional way. A sawmill A with no hard constraints will see an increased or decreased cost if another sawmill B adds constraints even if sawmill A does not. It is, therefore, an interesting aspect to study such effects as unintentional increased/decreased cost occurs when additional requirements are added. Moreover, if it is difficult to express added values, it would also be interesting to use a multi-objective formulation where cost can be evaluated against added quality satisfaction. Another area for more study is integrating the sorting done in the forest with three-dimensional (3D) sorting using scanners done in sawmills. Suppose some harvest areas transport unsorted sawlogs to a sawmill, then it may be enough to complement with sorting at some harvest areas in order to meet the properties required. Such an approach can efficiently combine log processing at harvesting areas, primary sorting together with secondary sorting at sawmills for further processing in wood manufacturing.

Acknowledgements

Thanks to John Arlinger, The Forestry Research Institute of Sweden, for his contribution with TimAn2 bucking simulations and Karin Westlund, The Forestry Research Institute of Sweden, for valuable discussions regarding the composition of the case.

Disclosure statement

No potential conflict of interest was reported by the authors.

References

Alzamora RM, Apiolaza LA. 2010. A hedonic approach to value Pinus radiata log traits for appearance-grade lumber production. Forest Sci. 56:281–289.

Amos F, Rönnqvist M, Gill G. 1997. Modelling the pooling problem at the New Zealand Refinery Company. J Oper Res Soc. 48:767–778.

Andersson G, Flisberg P, Liden B, Rönnqvist M. 2008. RuttOpt - a decision support system for routing of logging trucks. Can J Forest Res. 38:1784–1796.

Carlgren C-G, Carlsson D, Rönnqvist M. 2006. Log sorting in forest harvest areas integrated with transportation planning using back-hauling. Scand J Forest Res. 21:60–271.

Carlsson D, Rönnqvist M. 2007. Backhauling in forest transportation - models, methods and practical usage. Can J Forest Res. 37:2612–2623.

D'Amours S, Rönnqvist M, Weintraub A. 2008. Using operational research for supply chain planning in the forest product industry. INFOR. 46:47–64.

Epstein R, Morales R, Seron J, Weintraub A. 1999. Use of OR systems in the Chilean forest industries. Interfaces. 29:7–29.

Flisberg P, Frisk M, Rönnqvist, M,. 2014. Integrated harvest and logistic planning including road upgrading. Scand J Forest Res. 29:195–209.

Flisberg P, Rönnqvist M. 2007. Optimization based planning tools for routing of forwarders at harvest areas. Can J Forest Res. 37:2153–2163.

Forsberg M, Frisk M, Rönnqvist M. 2005. FlowOpt - a decision support tool for strategic and tactical transportation planning in forestry. Int J Forest Eng. 16:101–114.

Guajardo M, Rönnqvist M. 2015. Cost allocation in inventory pools of spare parts with service-differentiated demand classes. Int J Production Econ. 53:220–237.

Ivkovic M, Wu H, Kumar S. 2010. Bio-economic modelling as a method for determining economic weights for optimal multiple-trait tree selection. Silvae Genetica. 59:77–89.

Lidén B, Rönnqvist M. 2000. CustOpT - a model for customer optimized timber in the wood chain. In: Thorstenson A, Østergaard P, editors. Proceedings of the 12th Annual Conference for Nordic Researchers in Logistics, NOFOMA 2000; Jun 14−15; Aarhus (Denmark): Aarhus Business School. p. 421−441.

Maness TC, Adams DM. 1993. The combined optimization of log bucking and sawing strategies. Wood Fiber Sci. 23:296−314.

Misener R, Floudas CA. 2009. Advances for the pooling problem: modeling, global optimization, and computational studies - survey. Appl Comput Math. 8:3−22.

Misener R, Gounaris CE, Floudas CA. 2010. Mathematical modeling and global optimization of large-scale extended pooling problems with the (EPA) complex emissions constraints. Comput Chem Eng. 34:1432−1456.

Moberg L. 2006. Predicting knot properties of Picea abies and Pinus sylvestris from generic tree descriptors. Scand J Forest Res. 21:48−61.

Moberg L, Nordmark U. 2006. Predicting lumber volume and grade recovery for Scots pine stems using tree models and sawmill conversion simulation. Forest Prod J. 56:68−74.

Moberg L, Möller J, Sondell J. 2006. Automatic selection, bucking control, and sorting of sawlogs suitable for appearance-grade sawnwood for the furniture industry. N Z J Forest Sci. 36:216−231.

Möller JJ, Wilhelmsson L, Arlinger J, Moberg L, Sondell J. 2003. Automatic characterisation of wood properties by harvesters to improve customer orientated bucking and processing. In: Wide I, Båryd B, editors. Proceedings of the 2nd Forest Engineering Conference. Raw Material Utilization; 2003 May 12−15; Växjö (Sweden): Forestry Research Institute of Sweden. p. 64−76.

Næsset E, Gobakken T, Holmgren J, Hyyppä H, Hyyppä J, Maltamo M, Nilsson M, Olsson H, Persson Å, Söderman U. 2004. Laser scanning of forest resources: the nordic experience. Scand J Forest Res. 19:482−499.

Rönnqvist M, D'Amours S, Weintraub A, Jofre A, Gunn E, Haight RG Martell D, Murray AT, Romero C. 2015. Operational research challenges in forestry: 33 open problems. Ann Oper Res. 232:11−40.

Troncoso J, D'Amours S, Flisberg P, Rönnqvist M, Weintraub A. 2015. A mixed integer programming model to evaluate integrating strategies in the forest value chain - a case study in the Chilean forest industry. Can J Forest Res. 45:937−949.

Wilhelmsson L. 2006. Two models for predicting the number of annual rings in crossections of tree stems. Scand J Forest Res. 21:37−47.

Wilhelmsson L, Arlinger J, Spångberg K, Lundqvist SO, Grahn T, Hedenberg Ö, Olsson L. 2002. Models for predicting wood properties in stems of Picea abies and Pinus sylvestris in Sweden. Scand J Forest Res. 17:330−350.

Index